CLIMATE RESILIENT AGRICULTURE - STRATEGIES AND PERSPECTIVES

Edited by **Ch Srinivasa Rao, Arun K. Shanker** and **Chitra Shanker**

Climate Resilient Agriculture - Strategies and Perspectives
http://dx.doi.org/10.5772/intechopen.68382
Edited by Ch Srinivasa Rao, Arun K. Shanker and Chitra Shanker

Contributors

Angeles Calatayud, Consuelo Penella, Vanderlise Giongo, Alessandra Salviano, Francislene Angelotti, Tatiana Taura, Luiz Leite, Tony Cunha, Kizito Kwena, Fitih Ademe, Joseph Serge, Nezeghty Asmerom, Bernard Musana, Razaka Razakamiaramanana, Reuben Ruttoh, Hezron Mogaka, Assefa Dereje, Kifle Woldearegey, Anthony Esilaba, Rosemary Emongor, Upendra Sainju, Dhanej Thapa, Yuba Raj Subedi, Hemant Ojha, Zeynep Zaimoglu, Nafi Baytorun, Bruce McCarl, Panit Arunanondchai, Chengcheng Fei, José C Ramalho, Danielly Dubberstein, Weverton Rodrigues, José Semedo, Ana Rodrigues, Isabel Pais, António Leitão, Luiz Partelli, Eliemar Campostrini, Fernando Reboredo, Paula Scotti-Campos, Fernando Lidon, Ana Ribeiro-Barros, Fábio DaMatta, Andrés Cortés, Matthew W. Blair

Published by InTech

Janeza Trdine 9, 51000 Rijeka, Croatia

Publishing Process Manager Dajana Pemac
Technical Editor SPi Global
Cover InTech Design team

First published March, 2018
Printed in Croatia
Legal deposit, Croatia: National and University Library in Zagreb

Additional hard copies can be obtained from orders@intechopen.com

Climate Resilient Agriculture - Strategies and Perspectives , Edited by Ch Srinivasa Rao, Arun K. Shanker and Chitra Shanker
p. cm.
Print ISBN 978-953-51-3895-2
Online ISBN 978-953-51-3896-9

For EU product safety concerns:
IN TECH d.o.o., Prolaz Marije Krucifikse Kozulić 3, 51000 Rijeka, Croatia,
info@intechopen.com or visit our website at intechopen.com.

Meet the editors

Dr. Ch Srinivasa Rao is currently the director of ICAR-NAARM. He received his PhD degree from IARI, New Delhi, and postdoctoral degree from the Tel-Aviv University, Israel. He worked at IISS, IIPR, ICRISAT, and CRIDA. His expertise is in soil carbon sequestration, climatic change, contingency planning, and agriculture research management. He is a national coordinator of ICAR-flagship program on climatic change (NICRA). He was a technical chairman of the National Mission for Sustainable Agriculture and a climate change negotiator in the Indian delegation represented at UNFCCC, SBSTA, and COP meetings. He is an executive board member of IDDC, Cairo, and a member of Asian Carbon Network, Japan. He has published 250 research papers and 25 books and is a reviewer of 15 international journals. He is a research fellow of NAAS, ISSS, and ISPRD. He received prestigious awards from Hon'ble President of India, Hon'ble Prime Minister of India, ICAR, Indian Science Congress, IISS, NAAS, FAI, IARI, and ICRISAT.

Arun K. Shanker is serving as a principal scientist in Plant Physiology at the Indian Council of Agricultural Research (ICAR) of the Central Research Institute for Dryland Agriculture in Hyderabad, India. He is working with the ICAR as a full-time researcher since 1993 and completed his advanced degree in Crop Physiology while in service. He has been awarded the prestigious fellow of the Royal Society of Chemistry and fellow of the Royal Society of Biology. Presently, he is working on systems biology approach to study the mechanism of abiotic stress tolerance in crops. His main focus now is to unravel the mechanism of drought and heat stress response in plants to tackle the climatic change–related threats in agriculture.

Chitra Shanker is serving as a principal scientist in Agricultural Entomology at the Indian Council of Agricultural Research (ICAR) of the Indian Institute for Rice Research in Hyderabad, India. She has been working with the ICAR as a full-time researcher since 1995 and has completed her advanced degree in Agricultural Entomology from TNAU while in service. Earlier, she has worked on the insect interactions and biodiversity analysis in agroforestry. Her major work is based on invertebrate biodiversity in rice ecosystem and its potential for natural control of pests. She is now involved in the rice knowledge management portal in developing a farmer-friendly application for pest management decision support in rice, and her main focus now is on biocontrol agents for management of rice pests.

Contents

Preface

Climatic change denotes the long-term changes in climates including mean temperature and precipitation. Shifting weather patterns result in changing climate, which threatens the food production through high- and low-temperature regimes, increases the rainfall variability, raises the sea levels that contaminate coastal freshwater reserves, and increases the risk of flooding.

Agricultural productivity and climatic change are interrelated and interconnected processes, which take place on a global scale. Agriculture is particularly vulnerable to climatic change. Many forms of abiotic stresses, such as drought, heat, cold, frost, and flooding, and biotic stresses, such as disease and pest damage, that have the adverse effects on crop growth, development, and productivity can be exacerbated by climatic change in many regions. Although there will be an increase in yield of some crops in some regions of the world, the overall impacts of climatic change on agriculture are expected to be negative, threatening global food security. The importance of assessing the effects of global climatic changes on agriculture is a major task at hand for researchers to properly anticipate and adapt farming to maximize the agricultural production in a sustainable way. Variabilities in the local climates and the global climatic patterns have to be studied with equal importance in order to gain an understanding into the ways by which agriculture is being affected. The Earth's average surface temperature has increased by 0.83°C since 1880. Therefore, it is important to complete any assessment individually considering each local area.

One of the important methods to combat the ill effects of climatic change will be the natural resource management approach wherein there are a large number of options in soil, water, and nutrient management technologies that contribute to both adaptation and mitigation. Proven methods include *in-situ* moisture conservation, rainwater harvesting and recycling, efficient use of irrigation water, conservation agriculture, energy efficiency in agriculture, and use of poor-quality water. This, in addition to watershed management, which is now considered as a successful strategy, can be used as an adaptation measure.

Crop-based approaches form one of the important approaches in combating the climatic change, which include growing crops and varieties that fit into changed rainfall pattern; developing varieties with changed duration that can overwinter the transient effects of change; developing varieties of heat stress, drought, and submergence tolerance; and evolving varieties that respond positively in terms of growth and yield under high CO_2. The crop-based approach will be able to promote varieties with high fertilizer and radiation-use efficiency and also to promote novel crops and varieties that can tolerate coastal salinity and seawater inundation. In addition to this, the time-tested practice of intercropping can also be used as an effective strategy to combat the climatic change as it has the advantage under climatic variability and climatic change wherein if one crop can survive, the other crop may fail, thereby giving minimum assured returns for livelihood security. The productivity reductions expected under changed climatic scenarios are far more permanent than events of climatic variability. Crop improvement by using biotechnological and molecular advances involves identification and selection of traits that can increase tolerance to abiotic stresses and also enhance atmospheric CO_2 levels. Development of climate-ready varieties with traits, such as temperature, drought tolerance, and high yield of various important crops, is the need of the hour. The combination of these approaches is needed to evolve climate-ready cultivars. The genetic resources, mainly land races and wild relatives from the areas where there is an adverse climate, could serve as a source for biodiversity. The traits to look for when using biotechnology as a tool in both molecular breeding and transgenic programs are physiological traits, such as higher photosynthetic rate, better transpiration efficiency, lower respiration, lower nitrogen demand, lower tissue N, more stem/straw, less grain, and fast growth rate, and agronomic traits, such as input-use efficiency and shortened growing period.

The book focuses on various aspects of resilient agriculture and is a compilation of chapters that describe various strategies to combat the climatic change in agriculture with the ability to bounce back from damage caused by adverse environmental conditions. Effective natural resource management, innovative ideas, and strategic planning when used in combination can result in the development of adaptation and mitigation strategies. In addition, biodiversity at all levels—genes, species, and ecosystems—is an essential prerequisite for sustainable development in the face of a changing climate.

The answer to the challenge of countering the adverse effects of climatic change lies in resilience, and this can be effectively defined as the integration of adaptation, mitigation, and other practices in agriculture, which increases the ability of the system to respond to several climate-associated changes by resisting damage with an ability to recover rapidly. Such changes can comprise abiotic stresses, such as drought, flooding, heat and cold, unpredictable rainfall pattern, long dry spells, insect or pest population explosions, and other perceived threats caused by a changing climate. In short, it is the ability of the system to bounce back. Climate-resilient agriculture includes an inbuilt property in the system for the recognition of a threat that needs to be responded to in addition to the effectiveness of the response. The practice of resilient agriculture will essentially involve judicious and improved management of natural resources, namely, land, water, soil, and genetic resources through adoption of best bet practices.

This multiauthored edited compilation will attempt to put forth a comprehensive picture on most aspects of climate-resilient agriculture. An attempt is made here to synthesize and present information for developing strategies to combat the climatic change–related stress. In this book, we are presenting an ensemble of approaches that have a strong basic research background with an applied perceptive. The aim of the book is to cover the information that can bridge various areas of agricultural sciences, such as agronomy, soil science, plant breeding, and plant protection sciences so as to cater to a large audience who are working in climate-resilient agriculture.

Ch Srinivasa Rao
ICAR—National Academy of Agricultural Research Management (NAARM)
Rajendernagar, Hyderabad, India

Arun K. Shanker
ICAR—Central Research Institute for Dryland Agriculture (CRIDA)
Santoshnagar, Saidabad P.O, Hyderabad, India

Chitra Shanker
ICAR—Indian Institute for Rice Research (IIRR)
Rajendernagar, Hyderabad, India

Section 1

Adaptation and Mitigation

Chapter 1

Adaptation in Agriculture

Panit Arunanondchai, Chengcheng Fei and
Bruce A. McCarl

Additional information is available at the end of the chapter

http://dx.doi.org/10.5772/intechopen.72372

Abstract

Climate Change is continuing and happening faster than previously anticipated. Agriculture is vulnerable on a global scale and is currently adapting but will need to make further efforts in the future. Both public and private adaptation actions will need to occur, as certain potentially desirable adaptations are either not feasible or cost effective for private parties. Public action will play a crucial role in facilitating and supporting farmers to overcome barriers to adaptation and move toward a more sustainable and resilient agriculture. Here we discuss the sensitivity of agriculture to climate change and the adaptation strategies observed or potentially possible. We also discuss private and public roles in adaptation along with the constraints and barriers that limit adaptation. In addition, we discuss desirable factors to consider in adaptation project appraisal.

Keywords: climate change, agriculture, adaptation, project appraisal

1. Introduction

Climate Change (CC) is happening faster than previously anticipated and is altering agricultural conditions on a global scale. To reduce future climate risks, both mitigation and adaptation actions are likely necessary. While mitigation actions (control of greenhouse gas (GHG) emissions) reduce the future impact of CC, it takes time for such efforts to show significant effects and action has been slow to date. On the other hand, adaptation actions help reduce the negative effects of CC and can exploit opportunities [1]. Agriculture needs to adapt given the past and anticipated CC developments. Simply put agricultural adaptation is inevitable.

The evidence of that the climate is changing has grown dramatically during recent years. The IPCC mitigation report [2], shows the atmospheric concentration of GHG increased significantly with the pace increasing in recent times. According to IPCC [2], the atmospheric carbon dioxide (CO_2) concentration was 390.5 ppm in 2011, which is 40% greater than 1750 level, and exceeds 400 ppm today. Furthermore, counting other GHGs, the equivalent concentration is above 489 ppm [3]. Likewise, atmospheric nitrous oxide (N_2O) has increased by 20% since 1750 and atmospheric methane (CH_4) is 150% greater than before 1750 [2]. Greenhouse gas emissions are further projected to increase for many years. As a result, it is virtually certain that global mean surface temperatures will continue to rise. Global records show the Earth's surface temperature has been successively increasing with the three hottest observed conditions occurring in the last 3 years [4]. The numbers of cold days and nights have decreased, and we have seen increases in the number of warm days and nights and the length and frequency of heat waves [2]. Globally, precipitation has been increasing since 1901 with the spatial variability increasing and projections for large changes that differ across the planet [2].The precipitation change is not expected to be uniform with some regions projected to be drier.

All of these changes pose significant risks to agriculture. CC can affect crop yields, livestock production, water use, water supplies, and the incidence of weeds and pests among other items. CC adaptation actions are required to ensure a productive, profitable agriculture and in fact, adaptation is ongoing in the form of altered crop production locations, and planting/harvest timing along with other adaptations.

According to the IPCC [5], adaptation is the process of adjustment to actual or expected climate and its effects that in human systems, adaptation seeks to moderate or avoid harm or exploit beneficial opportunities. In addition, human adaptation actions may also occur in natural systems where actions are undertaken to facilitate "better" adjustments to the evolving climate. Such actions can benefit agriculture. For example, in a study in India, damages due to CC were predicted to be about 28 percent without taking adaptation, but 15–23% with it [6]. Moreover, in an agricultural context adaptation benefits also include reduction of income fluctuation, lessened agricultural vulnerability, and improvement in food security [5, 7–9].

A number of adaptation strategies, such as crop and livestock mix shifts, altered planting/harvesting dates, altered livestock stocking rates, and increased pesticide use, have been observed mostly implemented by farmers acting in their own best interests [10–14]. However, due to the available financial, technology, information human and physical capital not all adaptations can be privately implemented. Consequently, public parties will also play important roles in adaptation.

This paper addresses agricultural adaptation covering: (i) agricultural sensitivity and types of adaptation actions observed or projected; (ii) the roles that private and public parties play in adaptation implementation; (iii) the constraints and economic barriers that limit adaptation; and, (iv) the appraisal of adaptation projects.

2. Agriculture sensitivity and possible adaptation

CC and its drivers affect agriculture through a variety of changes, such as changes in temperature, precipitation, extreme event frequency and severity (e.g., floods, droughts, and heat waves) and increased atmospheric carbon dioxide [5]. Such factors alter crop and livestock yield, grass growth, livestock stocking rates, incidence of weeds and pests, and water usage among other items with geographically variability [5, 14–16]. Estimates in Beach et al. [17] show 30% projected increases in dryland corn yields in the southern US by year 2100, but little effects on the dryland corn yields in the western US and the Great Plains. Farmers have been observed to adapt on a regionally specific basis [5]. In this section, we review the nature of the effects and possible adaptations.

Finally, before starting we note that CC is not always as a negative factor but, depending on the region and situation, might also bring positive impacts to agriculture [14–20]. For example, Reilly et al. [15] presented results where moderate CC increases and the associated drivers can lead to increased cotton yield due to the effects of carbon dioxide and the drought tolerant nature of cotton. Furthermore, under such circumstances adaptation can be directed at increasing the positive aspects by say panting more of advantaged crops like cotton or increasing stocking rates for animals when grass growth is stimulated.

2.1. Crop yield

CC has been found to affect crop yield through alterations in temperature, precipitation, CO_2 level and extreme events [18, 20–23]. Lobell and Field [18] found that increasing precipitation had positive impacts on global crop yields, while the increasing monthly maximum and minimum temperature reduced yields resulting in a negative total effect. Adams et al. [21] found that yield effects vary across crops and regions in US with larger and more negative effects in the south but positive effects for some crops in the north. The Free Air CO_2 Enrichment (FACE) experimental data showed that the increase in atmospheric CO_2 concentrations increases the average yield of C3 crops (soybeans, cotton and wheat), but with little effects on the yield of C4 crops (corn and sorghum) except under drought (see the evidence reviewed in and developed by Attavanich and McCarl [20]). But Lobell and Field [18] argued that the carbon dioxide effects were minor and less than the effects of other CC factors. Moreover, Schlenker et al. [22] showed the incidence that extreme hot days (temperature over 34°C) was quite damaging to crop yields, and McCarl et al. [23] showed CC caused an increase in the variance of crop yield.

Farmers have adapted to the local climate since the beginning of farming activity. Today, they are coping with an increased pace of local CC. Crop timing changes, and crop-mix shifts are major observed adaptation strategies responding to crop yield changes. The United States Environmental Protection Agency (EPA) [12] illustrated that the average length of the US crop growing season increased by about 15 days between 1985 to 2015 with California and Arizona seasons increasing by almost 50 days. As a consequence planting dates have changed with corn and soybeans planted about 10 to 12 days earlier on average [24].

Several econometric and simulation approaches have been employed to study how farmers have been or could be adapting. Crop simulations show that adjustments to planting date and crop mix can greatly reduce the impact of CC with Aisabokhae, McCarl and Zhang [25] showing this to be the most valuable adaptation. In the U.S., the weighted centroid of major crops, such as wheat, corn and soybeans, were found to be moved to higher latitude plus a higher elevation and that the mix of crops in different locations changed [11, 23–27].

A multinomial choice model study by Seo and Mendelsohn [27] found that South American farmers in cooler locations preferred wheat and potatoes, while farmers in warmer locations planted more fruits and vegetables. They also found lower humidity pushed farmers to plant more maize and wheat [27]. Similar results were found by Park [28] using US data who found a continuum of adjustments with spring wheat dominating in cold regions, then as the climate warms corn and soybeans take over then cotton and sorghum and finally rice. Park [28] also found winter wheat and cotton were more selected in the dry locations, but soybeans were planted in the regions with more precipitation. Cho and McCarl [10] examined the impact of current and future climate on crop mixes over space in the U.S. and found that CC explained about 7–50% of the crop shift in latitude, 20–36% in longitude and 4–28% of that in elevation. Specifically, they showed that winter wheat production shifted northward and westward to cooler conditions in the Great Plains and spring wheat shifted east out of Oregon and Washington to higher altitudes and cooler temperatures in Idaho. Similarly, Fei, McCarl and Thayer [11] found the same pattern of adaptation in their modeling study. Moreover, Howden et al. [7] pointed out that CC also caused farmers to alter fertilization rates in order to maintain their crop quality.

2.2. Livestock production

CC affects livestock growth, diseases and mortality, animal reproduction rates, and quality of dairy products directly. It also has indirect effects via alterations in quality of feed crop and forage [29–31]. High temperature and humidity has been found to threaten the health, immune function, and mortality of livestock [30, 31]. Gaughan et al. [30] found elevated temperatures increased disease occurrence and mortality of new born calves plus decreased milk production. Mader et al. [29] showed that increased temperatures in hot areas decreased production of all livestock species. They also indicated that in the future the length of the feeding period would need to increase in hot southern U.S. regions, but shrink in the north [29]. Moreover, Hahn [31] showed the extreme heat waves in 1995 caused more than 4000 feedlot cattle deaths in Missouri and severe livestock performance losses in Illinois amounting to about a $28 million loss. Additionally, the number of animals dying in the heat wave in Europe 2006 and 2007 was large [30].

Livestock managers can adapt to CC through altered management, diversification of livestock varieties, alteration in livestock species and breeds, altered breeding practices, and modifying the timing of reproduction among other possibilities [7, 31–33] Hahn [31] recommended feedlot operators to use sprinklers for cooling, and to change the timing of handing and transporting particularly when the temperature humidity index (THI) is over 75. Rosenthal and Kurukulasuriya [32] provided evidence that diversification of livestock was effective in fighting against CC-related disease and pest outbreaks. Zhang, Hagerman and McCarl [33] found that summer heat stress was a significant factor for cattle breed selection in Texas with managers in regions with higher THI selecting more heat-tolerant cattle breeds (*Bos indicus*) than in other regions. A study in Africa showed that farmers changed both livestock species and mix with crops to adapt [34]. Specifically they found farmers increased reliance on livestock under hot and dry conditions, shifted to goats and sheep as opposed to cattle and chickens as temperature rose, and had more goats and chickens rather than cattle and sheep when precipitation increased [34]. Henry et al. [35] and Rowlinson [36] showed that changing breeding strategies could increase livestock tolerance to heat stress and diseases while also improving livestock reproduction. Better feeding practices, such as modification of diet composition, and changing feeding time, were found to improve the efficiency of livestock production [37].

Rosenthal et al. [32] found changing locations of livestock could help reduce soil erosion and improve moisture and nutrient retention, which in turn help adapt on the cropping side.

Herrero et al. [38] and Steinfeld et al. [39] showed that changes in crop-livestock system mix could improve efficiency by producing more food on less land using fewer resources, such as water. Mu, McCarl and Wein [13] found that increasing summer temperature plus an increased THI index caused adaptation in with the form of land switching from cropping to pasture, and stocking rates decreasing.

2.3. Water usage and supply

CC also affects agricultural water use and supply. CC alters plant evapotranspiration (ET) and thus water uptake [5]. Estimates by Adam et al. [21] show irrigated crops in the US south would need more water, but less in the north and mountain regions. Changes in precipitation, ET and crops planted have been found to affect water availability [5]. Even though the global average amount of precipitation did not show any significant changes in the recorded period (1991–2008), the observed spatial pattern and timing have changed, e.g. the frequency of heavy precipitation events increased and a downward trend of precipitation has been observed in Africa and South Asia [2, 40].

Substantial actions have been observed to adapt to water scarcity. In arid areas, where possible, irrigation and increased water storage have developed along with use of water saving technologies and drought-resilient crops [41]. Water management actions have also changed, with alterations in the amount and timing of irrigation, as have irrigation methods (with transitions from furrow to sprinkler irrigation [7, 42]). Moreover, water management strategies, such as smallholder irrigation development, rainwater harvesting, deficit irrigation and irrigation suspension are also possible as discussed in [43–47] .

2.4. Weed, pest and pathogens

CC can alter the pattern and incidence of weeds, pests and pathogens, in turn affecting crop and livestock performance and input cost [14, 47–49].

2.4.1. Weeds

Weeds compete with crops for sunlight, water, fertilizer and space, in turn reducing yields [47, 50]. CC and its drivers also affect weed growth, through temperature, precipitation, carbon dioxide concentrations, and other factors. In turn this affects the degree of competition with crops [47]. Moreover, CC is expected to shift the range of invasive weeds in turn introducing new weed issues in previously unaffected regions causing yield damages [51].

Herbicides and tillage are widely used weed control methods [49, 50]. Smith and Menalled [50] stated that integrated weed management, such as banding fertilizer near crop rows and applying it at the appropriate time, can help in adaptation, as well as can strategies such as reducing weed invasion and emergence, preventing weed reproduction, and minimizing the competition between weed and crops [50].

2.4.2. Pests and pathogens

CC affects the spread of pests affecting crops and livestock. It can enhance the spread of pests such as flies, ticks, and mosquitoes [52] plus increase disease transmission between hosts.

Mu et al. [53] found CC was associated with greater incidence of things like avian influenza. White et al. [54] found that CC led to increased Australian tick concentrations reducing animal weight by 18%. Howden et al. [7] discussed how wider use of integrated pest management can improve the effectiveness of pest control for livestock.

In terms of crops, CC induced changes in humility, precipitation and temperature have been found to alter the pattern of pathogens and their damages [49, 55]. Additionally more intense and more frequent rainfall events reduce the effectiveness of fungicides [49]. Chen and McCarl [14] showed that farmers adapted to CC induced increased pest incidence by increasing pesticide treatment costs. Wolfe et al. [49] indicated that one adaptation strategy is more frequent application of fungicides.

2.5. Climate-smart agriculture practices

Climate-Smart Agriculture (CSA) is a currently advocated adaptation approach. CSA is designed to jointly address food security and CC, enhancing CC adaptation and resilience [56]. CSA also addresses net GHG emissions. There are a number of CSA strategies and techniques that involve energy use, food storage, crop/livestock mix, water and soil management, and crop, livestock, forest, fisheries and aquaculture management [56].

3. Adaptation roles

Human adaptation through practices only occur if efforts are made to implement the practices. In many cases, farmers have been the implementers, for example altering crop rotations, changing crop mix, altering cultivars, changing the extent of tillage and revising timing of planting/harvesting. Those actions are undertaken since they are beneficial to the implementing individuals and enhance the performance of their farm business. However, some adaptation actions are so large (e.g. sea walls or research and development of climate adapted crop varieties) that farmers neither physically nor financially can individually invest in them. In addition, if they did invest in those items, those actions would benefit not only the farmer but also many others. That is, some adaptations have public-goods characteristics and following conventional economic theory will be underinvested in by private individuals [57]. This is where public action may be needed. More generally there are two forms of adaptation: (i) private (or autonomous in the literature) which is undertaken by individuals in their own best interest; (ii) public (or planned in the literature) that are adaptation actions undertaken by NGOs, or governments and designed to benefit broader elements of society (called public goods or actions to correct a market failure by economists). More details on public and private roles appear below.

CC adaptation to accommodate increases in severity or frequency of heat waves, floods, and climate related natural disasters is a great challenge. Public investments are needed to implement large and costly possibilities that benefit large segments of society or to facilitate private investment when it is limited by the state of technology, information, long-term nature of the investment or resources. Public sector adaptation efforts can take several courses: (i) they can provide incentives for private adaptation investments, such as subsidizing practices, releasing new technologies or providing low cost financing or grants [5]; (ii) they can set standards and regulations to require some degree of adaptation [58]; (iii) they can provide high-quality user-friendly information relative to adaptation needs, available strategies, and strategy implementation including technical assistance [59, 60]; (iv) they can localize information and practices to match specific regional conditions; (v) they can facilitate investment in long term adaptation capabilities (like a dam that lasts 50 years) [7, 28]; (vi) they can share in the risk and provide risk sharing mechanisms by providing insurance and possibly subsidizing it [5]; (vii) they can collaborate with the private sector in order to increase the efficiency of adaptation actions, such as carry out and providing results of R&D [61, 62]; (viii) they can identify and remove obstacles to adaptation, like distortions in input and output markets, trade barriers adverse subsidies and distorting insurance arrangements [63]; (ix) they can pursue policies that encourage adaptation investment in the context of the total development portfolio [57]; (x) they can also use policy levers like payments for environmental services; improved resource pricing; practice related subsidies, and taxes; alternative policies toward intellectual property rights; and direct actions or subsidies addressing adaptation enhancing research & technology development [5]. Studies by OECD on Innovation in Agriculture [64] and Innovation Strategy [65] provide supporting evidence that cooperation between public and private sectors improves the efficiency of public spending and induce more private firms to participate in adaptation. Those deciding on public priorities must consider whether the adaptations at hand would emerge through private actions either by farmers or supporting industries [66].

On the private side, many farmers are beginning to address climate change in their farm business planning acting in their own best interests. Farm management has incorporated climate risk into account by private parties and made changes as discussed in Section 2. In some cases, private parties can play a role in facilitating themselves into adaptation. Not surprisingly, the private sector is heavily involved in R&D of profitable crops and their products where private parties are willing to pay for the research outputs (like improved seed varieties) [66]. Private companies could also provide assistance for dissemination, creation and localization to farmers lacking information on climate change impacts and adaptation strategy type and performance [60]. Better information leads to development of private institutions to support agricultural adaptation in a variety of forms, such as better insurance, microloans, and other financial planning services. Nevertheless, there are numerous challenges when selecting appropriate adaptations. Cooperation between public and private sectors are necessary to resolve public good and market failure issues.

4. Adaptation characteristics

4.1. Adaptation constraints and economic barriers

While adaptation may be desirable, it certainly faces constraints. First, it is often constrained by available funds. Currently, on the public side there is a large gap between adaptation funding needs and actual funding. United Nations Framework Convention on Climate Change

(UNFCCC) predicted that about 28–67 billion USD funding is needed per year to adapt CC for all sectors [67]. More specifically, by year 2030 an annual estimate of global public funding needs for agriculture adaptation is about 2.3 billion U.S dollars per year [68]. However, current estimates place spending levels at around 1% of the need (FAO estimates 244 million USD for all sectors) has been provided for adaptation [69]. Moreover, adaptation is competing for funds with mitigation and non-climate investment (like education or military support or non-agricultural R&D) [5]. A balance across these three investments is required [5]. Wang and McCarl [70] find, compared to mitigation, that adaptation should get a larger investment share in the near term due to the cost inefficiency of mitigation, while when climate damage is large enough, mitigation should get more attention and investments due to increasing concentrations and damages.

Other main constraints include: (i) knowledge, and awareness, (ii) technology availability, (iii) physical and biological limits, (iv) economic and financial resources, (v) human capabilities and availability of the right types of people, (vi) social and cultural considerations, and (vii) governance and institutions [5].

There are also economic and individual behavioral barriers. These include transaction costs, information and adjustment costs, market failures, missing market, behavioral obstacles, ethics and distributional issues, coordination, government failures and uncertainty [5]. In addition, there are obstacles in the form of belief in whether CC is occurring and whether decision makers perceive needs for action.

4.2. Adaptation deficits and residual damages

Due to obstacles, belief and funding there is certainly an adaptation deficit, which is defined as "the gap between the current state of a system and a state that would minimize adverse impacts from existing climate conditions and variability" [5, 71]. Burton [71] argued and the funding gap shows that the adaptation deficit is growing. Huq et al. [72] argued this should be addressed by mainstreaming adaptation concerns in with development initiatives. It is also worthwhile noting that adaptation is unlikely to offset 100% of the damages with residual damages remaining. For example, it may be impossible to offset species extinction.

4.3. Maladaptation

Maladaptation is defined by IPCC as the case where adaptation actions lead to increased risk of adverse climate-related outcomes, increased vulnerability to climate change, or diminished welfare, now or in the future [5, 73]. The following two examples are considered. First, suppose a coastal city raises seawalls to eliminate the effects of sea level rise, and, in turn, the number of people and businesses located in the protected area increases. However, once future sea levels raise enough to overtop the seawall then more assets are vulnerable to CC. Second, suppose a major city with a river running through decides to divert water upstream as a flooding adaptation but the diversion results in increased flooding other places. However, despite these cases, we feel economically maladaptation should not always be prevented. Maladaptation can be rational, when overall net benefits of an action are positive and society can compensate losers [74].

5. Adaptation project appraisal

Given the emerging presence of adaptation concerns and in place funding mechanisms there will certainly be competitive funding situations where projects will need to be appraised and their costs and benefits compared. Like in the case of mitigation there are a number of adaptation project imperfections that merit consideration. Here, we outlined the some appraisal criteria covered by adaptation funding documents and the adaptation project imperfections as below.

According to Brann [75] and the Adaptation Funding Board [76, 77], current adaptation project appraisal focuses on strategic, operational and financing priorities, country eligibility, project eligibility, resource availability, implementing institution eligibility, implementing arrangement and performance monitoring.

However, in proposed projects, imperfections may arise relative to adaptation projects in terms of additionality, maladaptation, uncertainty, permanence, transactions costs, co-benefits, and their true adaptation nature. Each is dealt with below.

5.1. Additionality

In the adaptation arena, there are two definitions of the additionality concept. In one case, the concern is that the funding for adaptation represents an increase over funding for traditional development as opposed to a redirection of existing development funds [78]. The other concern is much like that in the Kyoto arena where one asks if by funding the project are adaptation benefits achieved that are above what would have happened in the absence of adaptation funding [79]. We will only deal with the later concern here.

Ideally, adaptation funding should stimulate additional action that reduces the detrimental effects of CC or exploits CC created opportunities. This implies the needs of some form of an additionality test that checks whether the adaptation would have happened anyhow. This could use tests like those under mitigation where the activity needs to be a documented money loser or something not already implemented in the region, as mentioned by Greiner and Michaelowa [80]. However, it may be desirable to use already implemented practices if they improve adaptation in a sub-region or among select parties where they are not now used.

5.2. Maladaptation

Some adaptation actions may lead to maladaptation where they may help reduce short-run CC effects but make adaptation worse for systems in other places or the future. This indicates a need for project applications to discuss impacts on other parties within the region or affected markets. It would also be desirable to cover the effects when the project is implemented on resources and economic activity in the vulnerable region including whether: (i) more economic activities will be stimulated in a region that becomes more vulnerable as CC proceeds and (ii) the current use of resources like depletable ground water precludes future use. In addition, one should evaluate whether the maladaptation may be acceptable or the benefits of the action are substantially greater than the costs of the maladaptation to the point whether potentially gainers could compensate losers.

5.3. Uncertainty

Yet another potential imperfection involves the degree of certainty manifest in the project effectiveness measures. An uncertain future climate and a possible lack of experience with project implementation and operation yield uncertainty. It would be desirable for projects to provide say a 90% confidence interval on benefits or an evaluation of performance under different future climates as an input to evaluation.

5.4. Permanence

Permanence involves the duration of the benefits from the project and embodies the assumption that the project benefits occur over a finite life, not forever. The degree of CC is expected to grow over time and thus the effectiveness of a given amount of adaptation expenditures is expected to fall as CC proceeds [81]. Therefore, it may be desirable for proposals to discuss activity life and its performance under escalating degrees of CC.

5.5. Transactions costs

Implementing and monitoring projects involves costs of passing funds to producers, insuring results, and observing/monitoring progress. Project applications might well cover the means for conveying funds, any brokerage fees involved and the methods/cost of insure the project operates as it is supposed to.

5.6. Co-benefits

Many possible adaptations have multidimensional implications including for example contributing to CC mitigation, improving current food security and correcting adaptation deficits. As such, project funders need to decide whether to evaluate or neglect such outcomes and the extent to which the funds flow into generating those co-benefits or to true adaptation. Project applications should layout not only adaptation benefits, but rather the entire spectrum [82, 83]. Nevertheless, evaluators must be cautioned as full consideration of co-benefits imposes a large burden in terms of identifying and quantifying co-benefits for all the projects at hand not just selected ones. In turn, evaluators will need to develop an approach to valuing the non-adaptation benefits and expenditures associated with the non- adaptation items stimulated by the project. However, Elbakidze and McCarl [84] argued that when appraising mitigation programs, the co-benefits should perhaps be neglected because co-benefits raises more uncertainties and burden to evaluate and qualify them across all projects, which should be considered here as well.

5.7. Is it really adaptation

The final issue is one of where the proposed activity is really adaptation. Lobell [19] argued that many suggested "adaptations" did not effectively reduce vulnerability under a changing climate. He argues that the benefits of adaptation actions should rise with degree of CC but that most of the ones cited in IPCC documents do not achieve this. Moore et al. [85] did a meta-analysis on the issue and found that virtually all the agriculturally related adaptation benefits reviewed by IPCC in the latest assessment report yield benefits that are independent of the changes in temperature or precipitation. This means such strategies while likely beneficial do not improve adaptation to CC.

6. Conclusion

Coping with a changing climate is an increasingly important issue in agriculture and one that is likely to persist for many years. In order to improve performance, farmers will need to adapt. Some adaptation efforts will happen in association with farmers making the best management decisions for their operation. But a class of possible adaptations can (i) face barriers being too costly for individual implantation, (ii) be hindered by substantial resource limits or behavioral barriers that the individual cannot overcome or (iii) yield benefits that spread widely across society and are not captured by an implementing individual. In such cases, public-sector intervention may play a crucial role in actual implementation or in the facilitation of producer adaptation implementation. Lastly, in comparative assessment of adaptation actions, one must also collect and evaluate information not only on benefits and costs of proposed adaptations but also on their imperfections in terms of additionality, maladaptation, uncertainty, permanence, transactions costs, co-benefits and true adaptation effectiveness.

Author details

Panit Arunanondchai, Chengcheng Fei and Bruce A. McCarl*

*Address all correspondence to: brucemccarl@gmail.com

Department of Agricultural Economics, Texas A&M University, USA

References

[1] Mendelsohn R. The economics of adaptation to climate change in developing countries. Climate Change Economics. 2012;**03**:1250006. DOI: 10.1142/S2010007812500066

[2] IPCC. Climate Change 2013: The Physical Science Basis. Contribution of Working Group I to the Fifth Assessment Report of the Intergovernmental Panel on Climate Change. Cambridge, United Kingdom and New York, NY, USA: Cambridge University Press; 2013. p. 1535. DOI: 10.1017/CBO9781107415324

[3] US Department of Commerce N. NOAA/ESRL Global Monitoring Division - The NOAA Annual Greenhouse Gas Index (AGGI) [Internet]. 2017. Available from: https://www.esrl.noaa.gov/gmd/aggi/aggi.html [Accessed: 2017-08-31]

[4] NOAA National Centers for Environmental Information (NCEI). State of the Climate: Global Climate Report for July 2017 [Internet]. 2017. Available from: https://www.ncdc.noaa.gov/sotc/global/201707 [Accessed: 2017-09-1]

[5] IPCC. Climate Change 2014: Impacts, Adaptation, and Vulnerability. Part a: Global and Sectoral Aspects. Contribution of Working Group II to the Fifth Assessment Report of the Intergovernmental Panel on Climate Change. Cambridge, United Kingdom and New York, NY, USA: Cambridge University Press; 2014. p. 1132

[6] Mendelsohn RO, Dinar A. Climate change, agriculture, and developing countries: Does adaptation matter? World Bank Research Observer. 1999;**14**:277-293. DOI: 10.1093/wbro/14.2.277

[7] Howden SM, Soussana J-F, Tubiello FN, Chhetri N, Dunlop M, Meinke H. Adapting agriculture to climate change. Proceedings of the National Academy of Sciences. 2007;**104**:19691-19696. DOI: 10.1073/pnas.0701890104

[8] Butt TA, McCarl BA, Angerer J, Dyke PT, Stuth JW. The economic and food security implications of climate change in Mali. Climatic Change. 2005;**68**:355-378. DOI: 10.1007/s10584-005-6014-0

[9] Butt TA, McCarl BA, Kergna AO. Policies for reducing agricultural sector vulnerability to climate change in Mali. Climate Policy. 2006;**5**:583-598. DOI: 10.1080/14693062.2006.9685580

[10] Cho SJ, McCarl BA. Climate change influences on crop mix shifts in the United States. Scientific Reports. 2017;**7**:40845. DOI: 10.1038/srep40845

[11] Fei CJ, McCarl BA, Thayer AW. Estimating the impacts of climate change and potential adaptation strategies on cereal grains in the United States. Frontiers in Ecology and Evolution. 2017;**5**:62. DOI: 10.3389/fevo.2017.00062

[12] US EPA. Climate Change Indicators: Length of Growing Season [Internet]. 2016. Available from: https://www.epa.gov/climate-indicators/climate-change-indicators-length-growing-season [Accessed: 2017-08-22]

[13] JE M, McCarl BA, Wein AM. Adaptation to climate change: Changes in farmland use and stocking rate in the U.S. Mitigation and Adaptation Strategies for Global Change. 2013;**18**:713-730. DOI: 10.1007/s11027-012-9384-4

[14] Chen CC, McCarl BA. An investigation of the relationship between pesticide usage and climate change. Climatic Change. 2001;**50**:475-487. DOI: 10.1023/A:1010655503471

[15] Reilly JM, Tubiello F, McCarl BA, Abler D, Darwin R, Fuglie K, et al. U.S. agriculture and climate change: New results. Climatic Change. 2003;**57**:43-67. DOI: 10.1023/A:1022103315424

[16] IPCC. Climate Change 2014: Synthesis Report. Contribution of Working Groups I, II and III to the Fifth Assessment Report of the Intergovernmental Panel on Climate Change. IPCC: Geneva, Switzerland; 2014. p. 151

[17] Beach RH, Cai Y, Thomson A, Zhang X, Jones R, McCarl BA, et al. Climate change impacts on US agriculture and forestry: Benefits of global climate stabilization. Environmental Research Letters. 2015;**10**:095004. DOI: 10.1088/1748-9326/10/9/095004

[18] Lobell DB, Field CB. Global scale climate–crop yield relationships and the impacts of recent warming. Environmental Research Letters. 2007;**2**:014002. DOI: 10.1088/1748-9326/2/1/014002

[19] Lobell DB. Climate change adaptation in crop production: Beware of illusions. Global Food Security. 2014;**3**:72-76. DOI: 10.1016/j.gfs.2014.05.002

[20] Attavanich W, McCarl BA. How is CO_2 affecting yields and technological progress? A statistical analysis. Climatic Change. 2014;**124**:747-762. DOI: 10.1007/s10584-014-1128-x

[21] Adams RM, Rosenzweig C, Peart R, Ritchie J, McCarl BA, Glyer JD, et al. Global climate change and US agriculture. Nature. 1990;**345**:219-224. DOI: 10.1038/345219a0

[22] Schlenker W, Hanemann WM, Fisher AC. The impact of global warming on U.S. agriculture: An econometric analysis of optimal growing conditions. The Review of Economics and Statistics. 2006;**88**:113-125. DOI: 10.1162/rest.2006.88.1.113

[23] McCarl BA, Villavicencio X, Wu XM. Climate change and future analysis: Is stationarity dying? American Journal of Agricultural Economics. 2008;**90**:1241-1247. DOI: 10.1111/j.1467-8276.2008.01211.x

[24] Sacks WJ, Kucharik CJ. Crop management and phenology trends in the U.S. Corn Belt: Impacts on yields, evapotranspiration and energy balance. Agricultural and Forest Meteorology. 2011;**151**:882-894. DOI: 10.1016/j.agrformet.2011.02.010

[25] Aisabokhae RA, McCarl BA, Zhang YW. Agricultural adaptation: Needs, findings and effects. In: Dinar A, Mendelsohn RO, editors. Handb. Clim. Change Agric. Cheltenham, UK and Northampton, MA, USA: Edward Elgar; 2011. pp. 327-341

[26] Adams RM, McCarl BA, Segerson K, Rosenzweig C, Bryant KJ, Dixon BL, et al. Economic effects of climate change on US agriculture. In: Mendelsohn R, Neumann JE, editors. Impact Clim. Change U. S. Econ. Cambridge, United Kingdom: Cambridge University Press; 1999. pp. 18-54

[27] Seo S, Mendelsohn ROA. Ricardian analysis of the impact of climate change on south American farms. Chilean Journal of Agricultural Research. 2008;**68**:69-79

[28] Park J. Essays on impacts of climate change on agricultural sector in the US [Thesis]. College Station, TX, US: Texas A & M University; 2012

[29] Mader TL, Frank KL, Harrington JA, Hahn GL, Nienaber JA. Potential climate change effects on warm-season livestock production in the Great Plains. Climatic Change. 2009;**97**:529-541. DOI: 10.1007/s10584-009-9615-1

[30] Gaughan J, Lacetera N, Valtorta SE, Khalifa HH, Hahn GL, Mader T. Response of domestic animals to climate challenges. In: Ebi K, Burton I, McGregor GR, editors. Biometeorol. Adapt. Clim. Var. Change. Vol. 1. Netherlands: Springer; 2009. pp. 131-170. DOI: 10.1007/978-1-4020-8921-3.ch7

[31] Hahn GL. Dynamic responses of cattle to thermal heat loads. Journal of Animal Science. 1997;**77**:10-20. DOI: 10.2527/1997.77suppl_210x

[32] Rosenthal S, Kurukulasuriya P. Climate change and agriculture : a review of impacts and adaptations [Internet]. 2013. Available from: http://documents.worldbank.org/

curated/en/757601468332407727/Climate-change-and-agriculture-a-review-of-impacts-and-adaptations [Accessed: 2017-08-24]

[33] Zhang YW, Hagerman AD, McCarl BA. Influence of climate factors on spatial distribution of Texas cattle breeds. Climatic Change. 2013;**118**:183-195. DOI: 10.1007/s10584-012-0642-y

[34] Seo SN, Mendelsohn RO, Dinar A, Kurukulasuriya P. Adapting to climate change mosaically: An analysis of African livestock management by agro-ecological zones. The B.E. Journal of Economic Analysis & Policy. 2009;**9**. DOI: 10.2202/1935-1682.1955

[35] Henry B, Charmley E, Eckard R, Gaughan JB, Hegarty R. Livestock production in a changing climate: Adaptation and mitigation research in Australia. Crop & Pasture Science. 2012;**63**:191-202. DOI: 10.1071/CP11169

[36] Rowlinson P. Adapting livestock production systems to climate change – Temperate zones. In: Rowlinson P, Steele M, Nefzaoui A, editors. Livest. Glob. Clim. Change. United Kingdom: Cambridge University Press; 2008. pp. 61-63

[37] Havlík P, Valin H, Mosnier A, Obersteiner M, Baker JS, Herrero M, et al. Crop productivity and the global livestock sector: Implications for land use change and greenhouse gas emissions. American Journal of Agricultural Economics. 2013;**95**:442-448. DOI: 10.1093/ajae/aas085

[38] Herrero M, Thornton PK, Notenbaert AMO, Msangi S, Wood S, Kruska RL, et al. Drivers of change in crop–livestock systems and their potential impacts on agro-ecosystems services and human wellbeing to 2030: A study commissioned by the CGIAR Systemwide Livestock Programme. ILRI; 2012

[39] Steinfeld H, Gerber P, Wassenaar T, Castel V, Rosales M, de Haan C. Livestock's long shadow [Internet]. 2006. Available from: http://www.fao.org/docrep/010/a0701e/a0701e00.HTM [Accessed: 2017-08-24]

[40] IPCC. Climate Change and Water. Technical Paper of the Intergovernmental Panel on Climate Change. Geneva: Intergovernmental Panel on Climate Change (IPCC) Secretariat; 2008. p. 210

[41] Barros VR et al.; IPCC. Climate Change 2014: Impacts, Adaptation, and Vulnerability - Part B: Regional Aspects - Contribution of Working Group II to the Fifth Assessment Report of the Intergovernmental Panel on Climate Change. Cambridge, United Kingdom and New York, NY, USA: Cambridge University Press; 2014. 688 p

[42] Ding J. Three essays on climate variability, water and agricultural production [Thesis]. College Station, TX, US: Texas A&M University; 2014

[43] Ngigi SN. Climate Change Adaptation Strategies: Water Resources Management Options for Smallholder Farming Systems in Sub-Saharan Africa. New York, NY: The Earth Institute at Columbia University; 2016

[44] Geerts S, Raes D. Deficit irrigation as an on-farm strategy to maximize crop water productivity in dry areas. Agricultural Water Management. 2009;**96**:1275-1284. DOI: 10.1016/j.agwat.2009.04.009

[45] Keplinger KO, McCarl BA, Chowdhury ME, Lacewell RD. Economic and hydrologie implications of suspending irrigation in dry years. Journal of Agricultural and Resource Economics. 1998;**23**:191-205

[46] Keplinger KO, McCarl BA. An evaluation of the 1997 Edwards aquifer irrigation suspension. Journal of the American Water Resources Association. 2000;**36**:889-901. DOI: 10.1111/j.1752-1688.2000.tb04314.x

[47] Patterson DT, Westbrook JK, Joyce RJV, Lingren PD, Rogasik J. Weeds, insects, and diseases. Climatic Change. 1999;**43**:711-727. DOI: 10.1023/A:1005549400875

[48] Gale P, Drew T, Phipps LP, David G, Wooldridge M. The effect of climate change on the occurrence and prevalence of livestock diseases in great Britain: A review. Journal of Applied Microbiology. 2009;**106**:1409-1423. DOI: 10.1111/j.1365-2672.2008.04036.x

[49] Wolfe DW, Ziska L, Petzoldt C, Seaman A, Chase L, Hayhoe K. Projected change in climate thresholds in the northeastern US: Implications for crops, pests, livestock, and farmers. Mitigation and Adaptation Strategies for Global Change. 2008;**13**:555-575. DOI: 10.1007/s11027-007-9125-2

[50] Smith RG, Menalled FD. Integrated Strategies for Managing Agricultural weeds: Making Cropping Systems Less Susceptible to Weed Colonization and Establishment [Internet]. 2012. Available from: http://msuextension.org/publications/AgandNaturalResources/MT200601AG.pdf [Accessed: 2017-05-22]

[51] Patterson DT. Weeds in a changing climate. Weed Science. 1995;**43**:685-700. DOI: 10.1017/S0043174500081832

[52] Thornton PK, Van de Steeg J, Notenbaert A, Herrero M. The impacts of climate change on livestock and livestock systems in developing countries: A review of what we know and what we need to know. Agricultural Systems. 2009;**101**:113-127. DOI: 10.1016/j.agsy.2009.05.002

[53] Mu JE, McCarl BA, XM W, Ward MP. Climate change and the risk of highly pathogenic avian influenza outbreaks in birds. British Journal of Environment and Climate Change. 2014;**4**:166-185. DOI: 10.9734/BJECC/2014/8888

[54] White N, Sutherst RW, Hall N, Whish-Wilson P. The vulnerability of the Australian beef industry to impacts of the cattle tick (Boophilus microplus) under climate change. Climatic Change. 2003;**61**:157-190. DOI: 10.1023/A:1026354712890

[55] Coakley SM, Scherm H, Chakraborty S. Climate change and plant disease management. Annual Review of Phytopathology. 1999;**37**:399-426. DOI: 10.1146/annurev.phyto.37.1.399

[56] Food and Agriculture Organization of the United Nations. Climate-smart agriculture: sourcebook. [Internet]. 2013. Available from: https://www.cabdirect.org/cabdirect/abstract/20153237305 [Accessed: 2017-08-18]

[57] Aakre S, Rübbelke DTG. Objectives of public economic policy and the adaptation to climate change. Journal of Environmental Planning and Management. 2010;**53**:767-791. DOI: 10.1080/09640568.2010.488116

[58] Harlow ED, McCarl BA. Climate change adaptation and disaster risk reduction for usaid's avansa agrikultura program in Timor-Leste 2015

[59] Wreford A, Moran D, Adger N. Climate change and agriculture: impacts, adaptation and mitigation [Internet]. 2010. Available from: http://www.oecd.org/tad/sustainable-agriculture/agriculture-and-climate-change.htm [Accessed: 2017-09-1]

[60] Vermeulen SJ, Campbell BM, Ingram JSI. Climate change and food systems. Annual Review of Environment and Resources. 2012;**37**:195-222. DOI: 10.1146/annurev-environ-020411-130608

[61] Organisation for Economic Co-Operation and Development (OECD). The OECD innovation strategy: getting a head start on tomorrow [Internet]. 2010. Available from: http://www.oecd-ilibrary.org/science-and-technology/the-oecd-innovation-strategy_9789264083479 -en [Accessed: 2017-09-1]

[62] OECD. Agricultural Innovation Systems: A Framework for Analysing the Role of the Government. Paris: OECD; 2013. p. 105

[63] Kimura S, Lankoski J, Antón J, Cattaneo A. A Comparative Study of Risk Management in Agriculture under Climate Change. 2012. DOI: 10.1787/5k94d6fx5bd8-en

[64] OECD. Agricultural Innovation Systems. Paris: Organisation for Economic Co-operation and Development; 2013

[65] OECD. The OECD Innovation Strategy. Paris: Organisation for Economic Co-operation and Development; 2010

[66] Naylor RL, Falcon WP, Goodman RM, Jahn MM, Sengooba T, Tefera H, et al. Biotechnology in the developing world: A case for increased investments in orphan crops. Food Policy. 2004;**29**:15-44. DOI: 10.1016/j.foodpol.2004.01.002

[67] UNFCCC. Investment and Financial Flows to Address Climate Change.Background paper on the analysis of existing and planned investment and financial flows relevant to the development of effective and appropriate international response to climate change. Bonn, Germany: United Nations Framework Convention on Climate Change (UNFCCC), UNFCCC Secretariat; 2007. 272 p

[68] McCarl BA. Adaptation Options for Agriculture, Forestry and Fisheries. A Report to the UNFCCC Secretariat Financial and Technical Support Division [Internet]. 2007. Available from: https://unfccc.int/files/cooperation_and_support/financial_mechanism/application/pdf/mccarl.pdf [Accessed: 2017-08-20]

[69] Elbehri A, Genest A, Burfisher ME. Global Action on Climate Change in Agriculture: Linkages to Food Security, Markets and Trade Policies in Developing Countries. Trade and Markets Division, Food and Agriculture Organization of the United Nations: Rome; 2011. p. 82

[70] Wang WW, McCarl BA. Temporal Investment in Climate Change Adaptation and Mitigation. Climate Change Economics. 2013;**04**:1350009

[71] Burton I. Climate Change and the Adaptation Deficit. Adaptation and Impacts Research Group, Meteorological Service of Canada, Environment Canada; 2004

[72] Huq S, Reid H, Konate M, Rahman A, Sokona Y, Crick F. Mainstreaming adaptation to climate change in least developed countries (LDCs). Climate Policy. 2004;**4**:25-43. DOI: 10.1080/14693062.2004.9685508

[73] Fan X, Fei CJ, McCarl BA. Adaptation: An agricultural challenge. Climate. 2017;**5**:56. DOI: 10.3390/cli5030056

[74] Arunanondchai P, Fei C, Fisher AC, McCarl BA, Wang WW, Yang YQ. How does climate change affect agriculture? In: Cramer G, Schmitz A, Paudel K, editors. Routledge Handb. Agric. Econ. Routledge; Forthcoming 2018

[75] Brann J. Climate Change Funding I: Briefing to the Africa-Caribbean-Pacific Group of States [Internet]. 2012. Available from: http://www.gcca.eu/sites/default/files/ACP/brann_af-ldcf-sccf-gef_acppresentation_14may_en_version_0.pdf

[76] Adaptation Fund Board (AFB). Adaptation Fund Project/Programme Review Criteria [Internet]. 2016. Available from: https://www.adaptation-fund.org/wp-content/uploads/2015/03/Review-Criteria-5.12.pdf

[77] Adaptation Fund Board (AFB). Operational Policies and Guidelines for Parties to Access Resources from the Adaptation Fund [Internet]. 2016. Available from: http://unfccc.int/files/adaptation/implementing_adaptation/adaptation_funding_interface/application/pdf/afbguide.pdf

[78] Lemos CM, Boyd E. The politics of adaptation across scales: The implications of additionality to policy choice and development. In: Boykoff MT, editor. Polit. Clim. Change Surv. Vol. Politics of. Series. London: Taylor & Francis; 2010

[79] United Nations Framework Convention on Climate Change. Kyoto Protocol to the United Nations Framework Convention on Climate Change [Internet]. 1998. Available from: https://unfccc.int/resource/docs/convkp/kpeng.html [Accessed: 2017-04-16]

[80] Greiner S, Michaelowa A. Defining investment additionality for CDM projects—Practical approaches. Energy Policy. 2003;**31**:1007-1015. DOI: 10.1016/S0301-4215(02)00142-8

[81] Parry M, Arnell N, Berry P, Dodman D, Fankhauser S, editors. Assessing the Costs of Adaptation to Climate Change: A Review of the UNFCCC and Other Recent Estimates; Martin Parry. London: International institute for environment and development (IIED); 2009. p. 111

[82] Brouwer R, van Ek R. Integrated ecological, economic and social impact assessment of alternative flood control policies in the Netherlands. Ecological Economics. 2004;**50**:1-21. DOI: 10.1016/j.ecolecon.2004.01.020

[83] Viguié V, Hallegatte S. Trade-offs and synergies in urban climate policies. Nature Climate Change. 2012;**2**:334-337. DOI: 10.1038/nclimate1434

[84] Elbakidze L, McCarl BA. Sequestration offsets versus direct emission reductions: Consideration of environmental co-effects. Ecological Economics. 2007;**60**:564-571

[85] Moore FC, Baldos ULC, Hertel TW, Diaz D. Welfare Changes from Climate Change Impacts on the Agricultural Sector: New Damage Functions from Over 1000 Yield Studies [Internet]. 2017. Available from: https://www.gtap.agecon.purdue.edu/resources/download/8367.pdf [Accessed: 2017-08-29]

Chapter 2

Climate Adaptive Agricultural Innovation in Nepal: Prospects and Challenges

Dhanej Thapa, Yuba Raj Subedi and Hemant Ojha

Additional information is available at the end of the chapter

http://dx.doi.org/10.5772/intechopen.72371

Abstract

It is evident that agriculture of Nepal is the most sensitive sector to climate change, and adaptations are essential for protecting the livelihood of rural poor farmers and ensuring their food security. Farmers are adopting different practices with climate awareness and sometime with spontaneity. In this chapter, we examined whether these climate change adaptation responses are adaptive and innovative and take into consideration existing and projected climate change and variability. Based on the review of adaptation theories and innovation approaches, researchers primarily draw a climate-adaptive agricultural innovation framework. We looked at empirically captured adaptation practices and analyzed their climate adaptive nature based on productivity, resilience, and equity. This study blends both qualitative and quantitative methods—combining case study of rice-wheat system with quantitative survey from four different regions of Terai, Nepal. The study shows that while agricultural system demonstrates a number of practices that contribute to adaptation, there are fundamental institutional, technological, and policy challenges that restrict the prospect of agricultural innovation required to adapt to changing climate.

Keywords: climatic stresses, agricultural adaptations, rice-wheat system, innovations

1. Introduction

1.1. Agriculture in climate change context

Agriculture is the most sensitive sector to climate change [1], as it can be affected by change in temperature, radiation, rainfall, soil moisture, and carbon dioxide (CO_2) with complex relationships [2]. Different climate change impact and sensitivity analysis conducted in global and regional levels have suggested decline in global food production [3–5]. However, climate change impact in agriculture is not uniform to all regions [6]. South Asian agriculture suffers more as three-fifth of the cropped area is rainfed; annual success of the monsoon determines well-being of millions of farmers [7].

Nepal, among the South Asia, is highly vulnerable and is ranked as a fourth most climate vulnerable country in the world [8]. Its fragile agro-ecology, flood-prone Terai region, weak infrastructural status, and poor economic condition put country to the vulnerable state [9]. It is estimated that the environmental income contributes to 50% of national GDP with agriculture, forestry, and fisheries [10]. Agriculture, which is largest contributor to Gross Domestic Product (GDP) of Nepal, is highly climate sensitive since only 53% of arable lands have year round irrigation facility [11].

Different studies conducted in Nepal suggest that farmers are suffering with climatic stresses of erratic nature of the monsoon rainfall pattern and climate extreme events [12–15]. Monsoon rainfall, which is more than 75% of total rainfall, determines the overall crop production of Nepal. Some time it has detrimental effects with increased precipitation during the monsoon with floods in lowland of Nepal. It is found to be an increased mean annual precipitation; with a decrease of rainy days with increased number of high-intensity rainfall events. It is recorded in most of the metrological stations of Nepal [16]. This scenario of rainfall intensity indicates more weather-related disasters, such as floods and landslides in future. Some studies show that flood incidence increased in Nepalese context has sharply declined crop production [17], severely affecting the life and livelihoods of the people in the entire Terai (lowland) region of Nepal. Every year, large sections of agricultural land are washed away by floods, and also degrading the land making it unsuitable for cropping. A research conducted by ICIMOD on the impact of climate change on water resources in the Himalayas shows that flooding will likely increase due to longer and more erratic monsoons, more intense rainfall events, and snowmelt, while droughts will increase due to glacier loss and changes in precipitation variability [18].

It is crucial for the Nepalese agriculture to adapt with climate change, since more than 66.5% of populations are primarily depended on agriculture. However, adapting to changing climate conditions is a huge challenge. This is particularly so for agriculture, which is generally well-adapted to mean or average conditions, but is susceptible to irregular or extreme conditions. More frequent droughts, floods, and deviations from "normal" growing season conditions, long-term changes in mean conditions, such as cumulative heat and timing of frosts, will have negative implications on agriculture [19, 20]. It is therefore necessary to understand the existing agricultural adaptation practices and critically analyze climate adaptability with existing and projected climate change and variability.

1.2. Concepts of climate change adaptation

According to dictionaries, adapt means to make suitable by altering. According to the Third Assessment Report of IPCC of 2001 [17], adaptation has been defined as *adjustment* in ecological, social, or economic systems in response to actual or expected climatic stimuli and their effects or impacts. This term refers to changes in processes, practices, or structures to moderate or offset potential damages or to take advantage of opportunities associated with changes in climate.

In agriculture, adaptation can be differentiated in the agro-biological system and human system. In agro-biological system, adaptation is always reactive, i.e., water stresses to rice shows lower rate of transpiration to minimize water stress; whereas, in human systems, it can also be anticipatory. Based on spontaneity of adaptation, it can be autonomous or planned [21]. Reactive and autonomous responses of human and natural systems do not necessary minimize short-term agricultural losses. The ecological, social, and economic costs of relying on reactive and autonomous adaptation to the cumulative effects of climate change are substantial. Also, short-term strategies taken by farmers and private sectors may have negative affect to long-term environmental damage. Adaptation to non-climatic stresses, however, influences adaptation to climate change, because it may increase the risk on mal-adaptation and thus reduce the resilience of a system to subsequence changes in climatic conditions [22]. So many of these costs can be avoided through planned and anticipatory adaptation. IPCC [19] report suggests that appropriately designed many adaptation strategies could provide multiple benefits in the medium and longer terms.

1.3. Climate change adaptation in agriculture of Nepal

In order to strengthen planned adaptation, Nepal has developed different policies and setup institutions. The Climate Change Policy (2011) and National Adaptation Programme of Action (NAPA) are the principal national-level policy and planning documents on climate change. To implement NAPA priorities at the local level, the government initiated of Local Adaptation Plans of Action (LAPAs) which involve the integration of top-down and bottom-up approaches to mainstream adaptation into planning from the local to the national level. Guidelines for LAPAs are included in the National Framework on Local Adaptation Plans for Action (Ministry of Environment, 22 November 2011), which provides the framework for the NAPA (and the Climate Change Policy 2011) to meet its mandatory provisions to disburse at least 80% of the available budget for the implementation of adaptation and climate change activities at the local level.

Ministry of Agriculture and Development (MOAD) has formulated Agriculture Development Strategy (ADS) 2015, which has a key component of biodiversity conservation and climate change adaptation and mitigation by (i) support the Local Adaptation Plan for Action (LAPA), (ii) scale up the interventions on soil conservation and watershed management including measures to promote adoption of sloping Agriculture Land Technology (SALT), and crop management practices (crop rotation, tillage, etc.), (iii) develop with policy decision, implement, and scale up schemes related to payment of environmental services including carbon sequestrations, and (iv) promote use of alternative/renewable energy and energy saving scheme among the local forestry groups. Recently, Ministry of Population and Environment has been in the process of preparing National Adaptation Plan (NAP) building on NAPA. While NAPA was immediate and short-term responses, NAP has planned to identify vulnerability and adaptation response for medium- (by 2030) and long-term (by 2050).

However, there seems still gaps and lack of understanding on how climate change impacts and adaptation needs will be identified and monitored in medium- and long-term. Even though many policies and plans have been formulated in favor of agriculture, but poorly executed within current institutional system. It is still a challenge for Nepal to comply its adaptation strategies with the existing and future climate change impacts and its uncertainty. With a wide micro-climatic variations and poorly managed limited hydro-metrological stations, weather-smart agricultural interventions, and crop-yield forecasting have a long way to go. In most cases, farmers are autonomously adapting with perceived climate risk in the local context. Nepal is still facing a challenge to identify and catalyze adaptive innovations, which can make agriculture resilience to the future climate change impact and associated uncertainty.

1.4. Conceptual framework of study

No matter what climate change adaptation options are and how they are determined, it should be climate adaptive with existing and future climate change. Accomplishing this task requires analysis of adaptation of farming and food systems from multiple lenses and approaches. Agricultural innovation system research has provided much useful information on the nature and dynamics of agricultural production systems and their responses to climatic and non-climatic stimuli. It characterizes agriculture as a complex system, within which changes are driven by the joint effects of economic, environmental, political, and social forces [23, 24]. This approach emphasizes the need to move from linear technology transfer model to more complex, process-based, interactive, and systems-oriented view of agricultural change and innovation.

In this chapter, we combine ideas on climate change adaptation with agricultural innovation and conceptualize climate adaptive agricultural innovation to understand and analyze adaptive responses of farmers and the local actors in relation to different stimuli from climate change and other socio-economic drivers. In understanding climate adaptive agriculture innovations, how farmers perceive risks and how they respond is certainly important, but how other actors and institutions mediate and shape even the farmers' perception and capacity is equally, and perhaps more, important in certain situations. Different studies have shown that decisions involving changes in agriculture are made at different levels by different agricultural actors that are interrelated with each other to form new pattern of agricultural system. In agricultural systems, farmers are the key actors, but they are not the autonomous agents. They are connected with traders, extension agents, fellow farmers, seed suppliers, and government regulators. The nature of farmer action is partly what farmer knows and partly what other actors advise or entice. In this research, we aim to explore the nature of such relations in relation to different innovations (**Figure 1**).

Our presumption is that farmers and agricultural stakeholders have been adopting their practices in relation to climate change and other drivers. But, due to the longer-term changes involved in climate regime, interventions based on local perceptions may not be fully adaptive

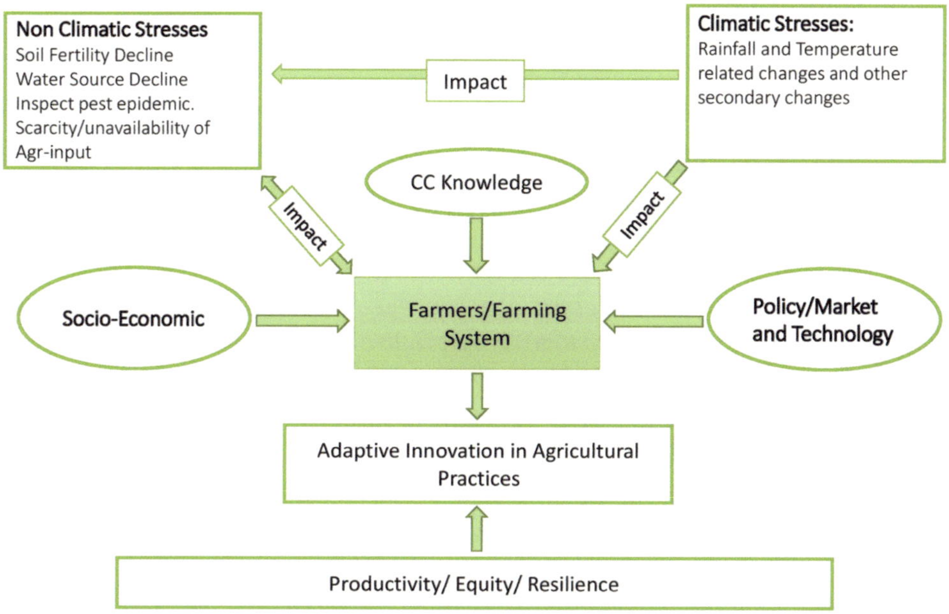

Figure 1. Conceptual framework of study.

to current and future potential risks. Establishing linkages among drivers, climate risks, and adaptation processes with respect to specific risks and opportunities can generate important insights into how climate change adaptation policy approach can be reframed to enable such adaptive practices in agriculture. Climate adaptive innovation practices are being been evaluated through farm productivity, resilience and equity; however farm and farmers' resilience is center of all.

2. Methodology

Both qualitative and quantitative inquiry approach has been taken for this study. Qualitative Inquiry approach has been taken to gather human experiences as a subjective experience, in different social context, and in historical time [25]. It is more to uncover knowledge about how farmers think and feel about different climatic stresses and their responses in local circumstances in which they find themselves.

Based on review of adaptation theories and innovation approaches, researchers primarily draw climate adaptive agricultural innovation framework. We look more at how the key actors like farmers, service providers, and regulators perceive, learn, and respond to diverse impacts of climatic change and variability on agriculture. Majorly, primary data are gathered from various studies conducted by the authors, including the one conducted with support from Climate Change Agriculture and Food Security (CCAFS), titled "Climate Adaptive Innovation: A Study of Agricultural Adaptation and Innovations in the Indo-Gangetic Plains, South Asia".

This study blends both qualitative and quantitative methods—combining a detailed case study of rice-wheat system with quantitative survey from four different CCAFS Block[1] (CB) of Terai, Nepal.

We purposively selected four districts (Sunsari, Rupandehi, Banke, and Kanchanpur) as our research site. Among five CBs of Nepal, only Sarlahi was not included mainly because of practical feasibility conducting the field work due to security situation at particular time. CBs[2] of Nepal are representative sites of Indo-Gangetic Plains in Nepal popularly known as Terai land [26]. The rice-wheat system is a common cropping practice in southern lowland Terai region, the bread basket for entire country. The two crops in the rice-wheat system together contribute more than 72% of total cereal production of Nepal [27]. Rice-wheat system is largely practiced in low land areas due to abundant fertile alluvial soils.

In order to conduct qualitative studies, Rupandehi district is selected, which is a lowland Terai of west-central Nepal. In the district, we purposively selected Hattibangai village development committee (VDC) for our detailed case study—considering the dynamic responses of agricultural stakeholders in the area. The VDC is located at the bank of Tinau River and is close to a major town in Nepal called Bhairhwa (Siddhartha Municipality). It is accessible by a black-topped road that crosses the village, and all the nine wards of the VDC are also connected to gravel roads. The community is heterogeneous comprising mixed caste groups and indigenous groups and new migrants. There are altogether 1076 households in the VDC spread out in the nine wards. Agriculture is either rainfed or irrigated through deep tube wells. The average land holding is 0.91 ha (**Figure 2**).

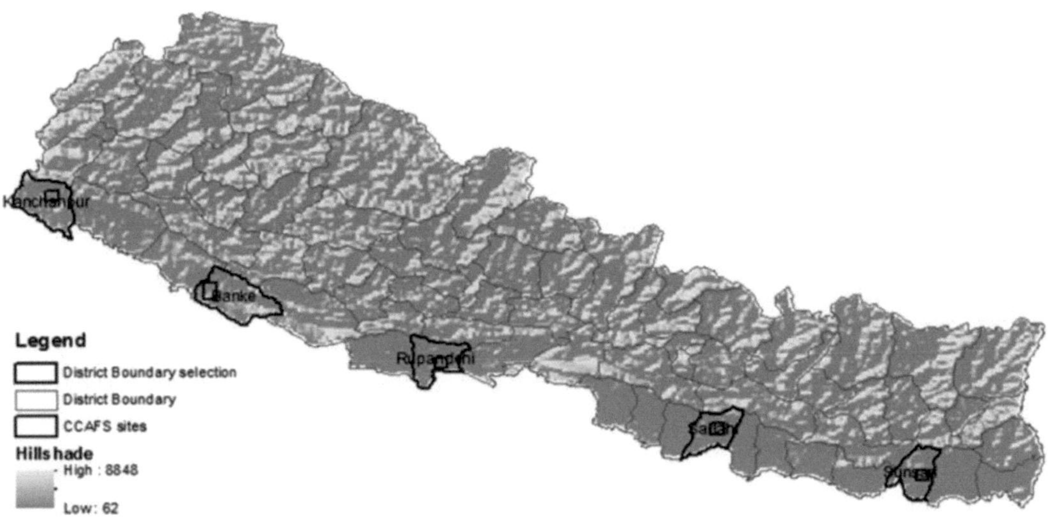

Figure 2. Map of Nepal showing CCAFS blocks.

[1]Climate Change Agriculture and Food Security (CCAFS).

[2]CCAFS blocks are 10 × 10 sq. m areas, where baseline survey has already been conducted by CGIAR Research Programme on Climate Change Agriculture and Food Security.

http://dx.doi.org/10.5772/intechopen.72371

Primary data were collected through different tools. Household survey was carried out in all four CCAFS block. Where as key informant interviews (KIIs) at community and district levels and focus group discussions (FGDs) with women and poor farmers, and household focused case-lets were done only in CCAFS block of Rupandehi district. During discussion and survey, climate-related questions were asked in later stage so as to avoid bias answer from participants.

All together 80 households, 20 households from each CB were selected for sampling survey. To make households more representatives in household's survey, three villages of CBs were selected using the map and list of villages from the CCAFS baseline data. Three villages of 200 hhs each located in 1 km of North-West corner of block, closest to center of block, and within 1 km of SE corner of block were taken for survey. With help of key informants in village, list of households were categorized into (a) big farmers and (b) small-marginal (land-poor) farmers. Among of them, sample of three big farmers and three small-marginal farmers were randomly selected from two corners of the block, whereas a sample of four big farmers and four small farmers were selected from the center.

3. Findings and discussion

3.1. Agricultural system changes and innovation

There are continuous changes observed in agriculture of Terai, Nepal. In dominant rice-wheat system, farmers have introduced multiple changes. Most of the respondent farmers were smallholders and were practicing subsistence agriculture followed by selling marketable surplus agri-products. Leasehold or share-cropping was very common to meet the need of sufficient food. Besides, agriculture was not only basis of livelihood; in fact, remittance played significant role in household annual income.

In recent years, farmers have either shifted to different agricultural practices or adopted new crop management practices. Nearly, 60% of respondent farmers shifted to new agricultural commodities. Vegetable cultivation was a chief source of new income sources to farmers. In most cases, winter-wheat crops were replaced by vegetable crops. About 50% of the respondent farmers of Banke district and 35% of the Kanchanpur district added vegetables. However, only 15% of respondents of Sunsari and 5% of Rupandehi district added vegetable crops in their farm. Around 5% of respondent of Kanchanpur district shifted to poultry production. A huge shift in banana farming was also observed in Rupandehi and Sunsari district. More than 15% respondent farmers of Rupandehi and 10% of Sunsari switched to banana (**Table 1**).

Most of them reported that shifting to new crops was more for higher income and for market security of these crops. It was noticed that most of the smallholder farmers had gone to such new changes. Smallholder famers did not prefer to cultivate rice-wheat crops.

One of the smallholder farmers of Rupandehi told us *"In wheat production I used to have a return of only NPR 20-25 thousand per 0.75 hectare of land, but in banana farming, it is possible to have a net income of more than NPR. 0.3 million."*

Block name	Old	New	Percentage of farmers
Banke	Rice-wheat	Vegetable	50
Kanchanpur	Rice-wheat	Vegetable	35
Rupandehi	Rice-wheat	Vegetable	5
Sunsari	Rice-wheat	Vegetable	15
Rupandehi	Rice-wheat	Banana	15
Sunsari	Rice-wheat	Banana	10
Kanchanpur	Rice-wheat	Poultry	5
Rupandehi	Rice-wheat	Pond fish	5

Field survey 2012.

Table 1. Households changing agricultural component/commodities over the 5 years.

Those farmers continuing similar crops have changed their crop management practices. However, time-related changes (changing time of agronomic management practices) was very less as compared to technological changes of crop management practices. Around 66% of respondents adopted new technology in land preparation. Farmers started to use tractors for land preparation. After decline of bullocks, use of tractors was common among farmers. However, introduction of tractor is not new in Nepal; a large number of farmers were also using tractors previously. In addition, farmers are also using new machines like rotavators[3] for land preparation during winter season for wheat crops. This has minimized the multiple tillage practices. Similarly, use of rotavators reduced the problem of water stress problem for low soil moisture by multiple tillage practices. More than 81% respondent farmers adopted new varieties in similar crops. In Hattibangai area of Rupandehi, more than 80% of farmers were using Indian hybrid rice (Gorakhnath) and Nepali drought tolerant Radh-4 rice varieties. Around 61% respondent farmers changed sowing and transplanting technologies. There were also significant changes in post-harvest operations technologies and use of pesticides, whereas shift in irrigation and harvesting technology was done by 66 and 64% of farmers, respectively.

Most of the respondents changed time of agronomic practices due to use of tractors, rotavator, and zero tillage accessories and combine harvesters. Also, time-related changes are more determined by adoption of new short duration varieties. Currently, time and frequency of multiple land preparations have changed. Majority of the farmers were being fully dependent on these new mechanical tools for land preparation (**Table 2**).

A research conducted by Thapa [28] used logistic regression model to analyze influencing non-climatic factors to switch new crops. Among the factors, contact with government organizations, climate awareness, subsistence farming were insignificant; whereas, small landholding farmers were 3 times more likely to switch toward new crops. Farmers' motive to switch new crops was largely driven by market profit. Similarly, farmers consulting to agro-vet were

[3]Zero tillage accessories are used in tractor for sowing wheat seed in row, and rotavetors are used to prepare land.

Agriculture management practices	Changes in	Block name				Total
		Banke	Kanchanpur	Rupandehi	Sunsari	
1. Land/soil preparation	Time	25	50	55	25	39
	Technology	40	55	100	70	66
2. New/crops varieties	Time	25	65	65	75	58
	Technology	75	75	100	80	82
3. Sowing/transplanting	Time	11	75	35	80	51
	Technology	11	90	70	75	63
4. Purchase/use of pesticides	Time		100	30	50	50
	Technology	44	95	100	70	78
5. Irrigation	Time	70	100	65	50	71
	Technology	55	95	100	20	68
6. Harvesting	Time	60	74	100	45	70
	Technology	60	74	100	30	66
7. Use/purchase of new agricultural implements	Time	50	—	40	30	24
	Technology	50	30	65	60	51
8. Post harvest operations	Time	30	21	46	35	32
	Technology	20	40	90	50	50
9. Marketing	Time	20	5	—	—	7
	Technology	10	10	—	5	7

Field survey 2012.

Table 2. New changes in agriculture management practices (in percentage).

4.8 times more likely to switch toward new crops. Famers consulting agro-vet got exposure on benefit from new crops as well as technical inputs for growing new crops.

Within rice-wheat system, farmers have adopted new practices such as improved varieties of rice and wheat from India as well as some released and registered varieties from Nepal. Improved varieties or rice imported from India and adopted by farmers are, Sarju 52, Sarju-49, Sava Mansuli, and Gorakhnath Gold. Similarly, Nepali improved varieties of rice adopted by farmers of Rupandehi are Mansuli, Loktantra, Mithila, Radha-4, and Barkhe-2b [29]. For wheat crops, National Wheat Research Program (NWRP) Rupandehi has introduced different varieties of wheat *Bijay*, *Adhitya*, and *Brikuti* with such characteristics as early maturing, drought resistance, and tolerance to pests and diseases. They are also popular among farmers of Rupandehi district [30].

Similarly, fish farming, which gives a higher return in a shorter period than rice farming, has also been adopted by farmers as a response to the repeated problems of floods washing away the rice crop and of declining yields. As a result of this development, farmers' income has increased due to fish farming, but at the cost of a decreasing rice plantation area [29]. Farmers of Rupandehi have gone through series of agricultural experiments in their farms such as cash crops, vegetables, banana, fish, and other management practices [31]. A CCAFS sites in Rupandehi exhibit the highest levels of diversity in production, with over 50% of surveyed households producing more than eight different products [32]. However, such changes are shaped by different factors. In lowland with clay soils, farmers have shifted to fish pond where as in low sandy loam soils, farmers have shifted to banana crops.

If we see the household priorities and behavior, in terms of flexibility toward adaptive change and belief in investing in new innovations, smallholders' farmers are adopting labor intensive income oriented cash crops. Whereas, large holder famers managed labor scarcity, thereby increased use of new machines. There are other changes and adaptation responses such as (i) increasing attraction toward crop security program, (ii) changing their seeds in 2–3 years, (iii) crop rotation with legume crops, (iv) increasing trend of groups and co-operatives formation and private agro-vets.

3.2. Climate change realities and farmers' perceptions

However, all the changes in agriculture of Terai are not well informed by existing climate change. It seems more autonomous with certain climate stimuli and market. There exist huge gaps in understanding the climate adaptability of these changes and adaptation responses. Local people shared some experiences of the climate conditions, but most of the respondents were not aware about climate change. However, they have felt different climate extreme events and negative impact to agricultural production. Farmers were showing their direct concerns on precipitation rather than temperature-related issues. Rice farmers were worried about uncertain monsoon rainfall. Delayed and uncertain monsoon has left farmers in dilemma specially in raising the nursery bed.

One of the farmers of Gargatti village of Rupandehi remembered. *"It used to be delay in monsoon rainfall in the past years but this year it was early rain and we did not have rice seedlings for transplanting. Rainfall was continuous but we were unable to transplant rice"*. A Chief of District Agriculture Development Office (DADO) added *"monsoon came early in last year and very less precipitation was seen in August and September, drastic reduction in mid and late varieties like Sawa Mansuli and Sunaulo Sugandha varieties of rice"*.

While asking famers about their future risk perception, 49% farmers said drought as very serious risk in future as well and 55% farmers feel that somewhat serious.

In case of temperature related changes, only few of them linked to their current agricultural practices. Wheat farmers have felt that wheat-growing season is being delayed. Usually, the farmers used to seed wheat during first week of November. But now, they have shifted to the third week of November. Senior Scientist from National Wheat Research Program (NWRP) told us that generally November 15 is used to be the appropriate time for wheat sowing, but now wheat sown 10 days later gives more yield.

However, famers' perception on climate change and overall ranking taken from weighted mean showed six major climatic stresses perceived by farmers. Rainfall difficult to predict, drought, increasing summer temperature, monsoon starts earlier, winter has become cooler, and more rainfall intense rainfall during monsoon were ranked from one to six.

The analysis of 30 years climate data shows that there is evidence of climate change in the study site. Monsoon precipitation anomaly generated from 30 years monthly rainfall data from the nearest meteorological station (Bhairahawa Airport) highlighted the decreasing trend of precipitation in the area. This is linked with the farmers' perceptions on increasing drought condition and increasing summer temperature. Interestingly, drought was more emphasized by local leader farmers and district stakeholders and agro vets. However, there were floods in the years of 1981, 1984, 1989, 1998, and 2006.

However, our survey finding revealed drought as fifth important observation. This might be due to excess use of ground water for irrigation. Farmers were using their boring water for irrigation. Farmers in Hattibangai and Rupandehi areas have also perceived decline of water table. Farmers reported that deep-tube wells/boring sets need to go further depth for harvesting water. According to farmers' perception, water table has declined from 180 to 280 ft. A research conducted by Dahal et al. [33] suggested decline of water table in nearby areas. Even some of the Shallow Tube Wells (7.62 m) were dried during dry season (from April to June) in many cases in the vicinity of the Tinau River. However, drying of Shallow Tube Wells (STW) was linked with excess riverbed extraction of Tinau River. In the same paper, it was clearly argued that decline of water table of deeper STW depths from 28.96 to 36.58 m were not affected by the extraction in the Tinau River. In the case of Hattibangai and nearby VDCs, Bhairawa Lumbini Groundwater Project constructed number of deep boring sets for irrigation purpose. Excess of ground water harvesting by deep boring sets led to the decline of water table. In case of declining precipitation, water table failed to recharge as per required. So, we can conclude that decline of water table is more linked with decrease of precipitation.

The analysis of 30 years of temperature data shows a slight increase in minimum temperatures while maximum temperatures have remained the same [29]. While taking farmers perceptions, 49% respondent farmers' perceived decrease of winter temperature and 81% of farmers accepted negative effects due to chilling winter. However, climate data showed increase of minimum temperature. Few farmers' perceptions were matched with the climate, while others did not. While analyzing the perception of farmers on rainfall, around 42% of respondents reported that intensity of rainfall was increasing, while precipitation analysis showed a decline of rainfall during the last 10 years. However, if we look at the precipitation of Butwal station, sharp increase in rate of precipitation was observed [28].

Late monsoon and hotter winter were insignificant to explain influencing climatic factors to switch crops. Farmers experiencing drought were 5.5 times more likely to switch new crops, as compared to those who did not experience drought as a problem. Similarly, farmers experiencing early winter more severe were 8 times more likely to switch new crops [28].

Since the study was conducted in rainfed area, farmers significantly used tube wells for irrigation. However, it was more costly and farmers were demotivated for frequent use of tube wells. Similar with our research finding, another research of Gauchan and Gumma [34] showed that drought incidence has increased in Nepalese context. This study analyzed production and productivity by analyzing satellite images taken spatially and temporally on rice crops and drought years. This can be predicted that with increasing climate change, such situation can be further devastating. Amgain et al. [35] in their simulation study in Indian context indicated that increments in both maximum and minimum temperatures by 4°C decreases rice yield by 34% and wheat yield by 4% as compared to base scenario with current weather data. However, in Nepalese context, hot summer was significant, but farmers were 0.144 times less likely to switch crops as compared to that respondent who did not experience hot summer. These research findings suggest that farmers are changing cropping pattern as a short-term strategic actions to cope with problem of water.

It is obvious that farmers have experienced climate risk. Their perceptions on climate change and consequences were sometime matched with climate data analysis, whereas contradictory in many cases. There are strong gaps existing between the predicted level of climate change and the actual adaptive actions among both the farmers and other locally based agriculture stakeholders, suggesting the deficit of processes and institutions to facilitate adaptive innovations. By and large, climate science data still remains within the research institutions, not readily accessible to agricultural actors.

3.3. Adaptive innovation support system

Adaptive activity does not occur in institutional vacuum [36]; range of institutions from household to community to government systems affect choices of individual farmers. Several institutional changes are noticeable.

Extension policies supported the formation of cooperative groups, which enable to continue banana farming, rice-wheat crop intensification, and farm mechanization. Promoting private sector policies encouraged to establish agro-vets in local level, providing seeds/pesticides and technical information to vegetables growers. Generally, farmers contact to government organizations for new agricultural implements, whereas co-operatives and private agro vets for seeds and pesticides, respectively. Farmers are usually depending on local institutions like co-operative, VDC level government agricultural service centers. However, climate information was not shared by such institutions. Farmers have limited access to climate related information. Only 23% farmers were getting weather forecasts. Among of them, 36.3% of them were receiving from radio. 31.5% from community meetings, 25% farmers from "extension workers" and 14% from "neighbors." Whereas, the use of "mobile phone," "email/internet," and "newspaper" were relatively very low.

Farmers usually do not visit meso-level institutions located at district for climate agricultural information. Farmers' perceptions and responses toward climate change are more autonomous actions and lesser based on scientific knowledge of climate change gained from extension services. Still there are key barriers like poor access to mass media, and poor information flow on climate change. Farmers think for need of more community meetings and workshop facilitated by meso-level extension agencies rather than mobile SMS and other mass media. Agricultural extension system, which is largely within government, is also slow to adapt and communicate climate science to farmers, as there is still limited institutional priority accorded in processing and communicating the scientific knowledge.

Market is becoming an important driver to agricultural change. It is not just the market of agricultural inputs or outputs but also the opportunity costs or the relative value of substitutes for producing agricultural commodities. With the shut down of the sugar factory in Rupandehi district, it was a poor demand for sugarcane and so farmers switched to banana crops. Market has become a handy provider of inputs and even technical advices to farmers. Seed store/agro-vets are examples local agribusinesses, which largely influence decisions of farmers. Larger equipment such as tractors/other farm mechanization tools are also available in the small and medium size towns. These also make it possible for the small farmers to access the services of relatively costly equipment on per hour basis without a need to own. Besides, sometime farmers suffered with numerous problems regarding different issues of crop failure due to seed sterility, low market price, and unavailability of seeds and fertilizers. Adaptive innovations are not just steered through incentives or disincentives offered by policy or something that occur in response to the market based incentives.

Technologies are being introduced in farms with multiple interactions of market, extension and farmers network. Currently, Terai of Nepal has different levels of technological sophistication in agriculture. Government policies and market forces both promoted mechanization. Subsidies exist to enable farmers to have access to technologies. But, the production focus is not clearly articulated with longer-term adaptation to climate change. Some climate-friendly technologies introduced earlier are lesser practiced by farmers such as zero tillage, integrated pest management (IPM), and sustainable soil management (SSM). There is a lack of fundamental rethinking on the need to link technology with sustainability and resilience. In Rupandehi, zero tillage was introduced to farmers to sow wheat as soon as rice is harvested, and the crop matures before hot winds of spring shrivel undeveloped grains. However, farmers are more attracted toward rotavator, which have many negative implications to soil properties and moisture conservation.

Civic engagement compliments an important part [37, 38] in climate adaptive innovation. This includes farmer-to farmer co-operation, formation of associations for advancing the interests and concerns, and even lobbying with political leaders on policy issues affecting agriculture. There seems increasing number of farmers' organizations/network (formal/informal), especially groups of farmers. Hattbangai, Rupandehi has diverse groups—famer groups, women farmer groups, saving credit groups, few groups form by DDC/VDC and I/NGOs. They have introduced different agriculture practices and farmers were overwhelmingly positive to the opportunity they received in participating in various activities. More than 76% of respondent farmers found it useful. As the Nepal's political environment is becoming democratic, farmers have enjoyed more opportunities in organizing themselves. However, there was also a challenge to active civic engagement due to factional divisions among political parties.

Agriculture policy approach continues to be top-down and linear [39] while there is an increasing need for "a comprehensive and dynamic policy approach, covering a range of scales and issues" [40] in the context of climate change. The formulation of NAP-Agriculture and ADS 2015, which aimed to direct the climate change adaptation in agriculture, are more or less failed to consider ground realities and complexities of adaptation challenges in farm level. To enhance the adaptability of agriculture, short-term and farm-level adaptation actions of farmers should be made part of the efforts to secure long-term resilience of the entire agro-ecological systems in the localities. However, regulatory responses have remained mixed, but largely ignorant of the current and future effects of climate change on agriculture. In Rupandehi, there was a number of ground water projects, which are currently not functioning with decline of ground water table. There is continuous deepening of Tinau River, affecting irrigation system in long run.

3.4. Outcome of agricultural changes and innovation

Adaptive Innovation and equity: Current status of adaptive innovation practices show a number of challenges related to distributional outcomes (equity). In fundamental sense, adaptability of agricultural systems also depends on its ability to deliver equity and fairness. Agricultural changes in different farm practices were not always gender friendly. While assessing status of women's workload changed over the past 5 years, 59% of respondent farmers said that workload to women during agricultural field preparation has been decreased. However, in sowing crops, 51% of respondent farmers said that workload has been increased. With use of tractors, combine harvester, rotavators, and irrigation related used of boring sets supported the traditionally defined work of men. Whereas women-involved work such as intercultural operations and sowing of crops were less supported by current mechanization. In fact, in some cases, women suffered more. For example, women of those involved in banana farming invested their extra time and effort for fertilizer and pesticide applications. While on marketing, processing, and storage, we had mixed perceptions on increased, decreased, and no difference. However, slightly more percentage of farmers were saying that workload had decreased. Basically, decisions on these changes are done by male members who may lead to less adoption of women friendly technologies in sowing, transplanting, and post-harvest operations. More than 75% farmers told that farming decision was taken by male members (**Table 3**).

Activities	Decreased	Increased	No difference
1. Getting agri-inputs	26	18	36
2. Agricultural field preparation	39	17	24
3. Sowing	10	41	29
4. Transplanting	8	48	24
5. Inter-cultural operation	10	53	17
6. Harvesting	25	41	14
7. Processing	30	27	23
8. Storage	34	27	19
9. Marketing	31	14	35

Field survey 2012.

Table 3. Status of women's workload change over the past 5 years (in number).

A number of observations can be made with regard to the equity aspects. Most of the agricultural technology has favored large landholders, male members, and there is limited research attention to explore and develop pro-poor technology. In some instances, NGO support has allowed landless to have access to land and technology. But, there still remains a question on institutional sustainability, as there are no links with established and accountable system of local and national governments. Much of the technological innovations have not addressed the workload of women. Current farm mechanization failed to reduce burden of female farm worker.

Adaptive innovation and productivity: Farming systems are organized around small and fragmented land holding. In such situations, an adaptive action of farmers takes place in small pieces of land and shaped by motive of production maximization. There is little awareness among the meso-level players of agricultural systems on how adaptation and production innovations could and should be combined together. The question then is how short adaptation actions of farmers become part of long-term efficient production, such that short-term adaptive actions do not hamper the ability of the system to adapt to more intense and large scales shocks that are likely in the long run [40]. Some adapt practices such as crop rotation in rice-wheat crops with legumes and crop diversifications. Similarly, bio-intensification was done by 10% farmers with less priority. Only 23% of farmers carried out any resource conservation practices in farm, while none of the famers have conducted organic farming as a planned and systemic manner. Fallowing practice has also left by farmers. Only 16% farmers left fallow for scientific practice.

Current service delivery system is still dominated by state agencies and there is a lack of mechanisms to ensure other non-state institutions to be responsive to the needs of the farmers. For instance, Nepal's Agri-Input Corporation (AIC) had problems with farmers in regular supply of seeds and fertilizers. Famers often compelled to adopt other high-yielding varieties available in market; which are poorly tested local farm context. Famers are overwhelmingly using single crop varieties with aim to increased higher production lead to risk of crop failure. It requires going beyond "intensification" of agriculture, to link options for sustainability and resilience more seriously. Farmers' ability to engage in climate adaptive innovative practice is substantially shaped and mediated by the stakeholders operating at local and meso-levels. The capacity of household to cope with climate risks depends to some degree on the enabling environment of the community, and the adaptive capacity of the community is reflective of the resources and processes of the region [41]. But, everywhere such meso-level institutions demonstrate much less adaptive response than the farmers.

Adaptive innovation and resilience: Different farm level changes in agriculture are outcome of their transformative learning; triggered disorienting dilemmas to farmers. Social learning in agricultural and natural resource management has remained a key aspect to keep the system resilient. Learning can take place as anticipatory [42] or can occur as transformative change [43] following some crisis. In Rupandehi, a severe hailstone in 2000 damaged one variety (Bhrikuti) of wheat, compared to others (NL 297 and UP-343). Famers are still continuing NL 297 variety, which is discouraged by DADO. Farmer shifted to banana from sugarcane after sudden shut down of sugar factory lead to extra burden to female farmers. Similarly, farmers started to sow more than two to three varieties of rice after they suffered with a heavy insect infestation in one variety, whereas less in others. However, such transformative learning and knowledge are remained within small farm level.

In some cases, especially when farmers have access to services and information, farmers have resorted to adaptive and innovative practices—such as changing cropping patterns, technological changes. But again, such innovations lack backing by adaptation thinking informed by an analysis of current and future effects of climate change and variability. Farmers' perception toward climate is not part of their systemic analysis of long-term changes in climate. Farmers are aware of noticeable variation in weather patterns. Farmers linked their problems with hot summer days, erratic nature of monsoon rainfall and drought. However, there exist strong gaps between perceptions of climate change and the adaptive actions among both the farmers and local stakeholders, suggesting the deficit of processes and institutions to translate information into adaptive actions.

Adaptation planning and policy systems have followed traditional sectoral, administrative paths, and at times contributed to maladaptation. The nature of risk varies tremendously across sites and so are the capacities and resources of the actors. The case of water decline is more noteworthy from this point of view—as farmers are all set to maximize individual farm production, while there are little public concerns over the declining water quality and groundwater stock.

4. Conclusion

This study confirms that farmers are the active agent of change in agriculture. They experiment, introduce, and experiment different farm level changes with a motive to enhance their farm production, productivity, and sustaining their agrarian livelihood. Farmers have identified different adaptive and innovative responses. However, there is still a lack of framework to understand and catalyze adaptive responses in such a way that it is informed by long-term trends in climate change. There are fundamental institutional, technological, and policy challenges that restrict the prospect of agricultural innovation required to adapt to changing climate. Most of actors have considered private risks in the short run and predominance of market logic. This lacks attention to adaptive aspects—still focusing on productivity aspects, i.e., monoculture of few crops, high yield varieties (HYVs) face major risks, such as over harvesting of natural capital, e.g., ground water.

Farmers' perception toward climate is not part of their systemic analysis of long-term changes in climate. There exist strong gaps between perceptions of climate change and the adaptive actions among both the farmers and local stakeholders, suggesting the deficit of processes and institutions to translate information into adaptive actions. By and large, climate science data still remains within the research institutions, not readily accessible to agricultural actors. Agricultural extension system, which is largely within the government, is also slow to adapt and communicate climate science to farmers, as there is still limited institutional priority accorded in processing and communicating the scientific knowledge. In order to flow of climate information and alertness about climate change are essential. It is necessary to cautiously design effective climate information flow system to reached larger numbers farmers. For an agricultural innovation to be climate adaptive, it has to be informed by the expected change in the climate.

Farmers have responded to climatic and socio-economic drivers to agriculture, which involve a wide range of social, technological, political, environmental adjustments, often in association with a wide range of agricultural stakeholders in the region. But, these changes have not been internalized adequately by the institutions and policy systems. In overall, we can conclude that current agricultural system demonstrates a number of practices that contribute to adaptation, but it should be well informed by climate information and facilitated by institution and policy system that enables agricultural innovation climate adaptive to changing climate.

Acknowledgements

This chapter is based on research on "*Climate Adaptive Innovation: A Study of Agricultural Adaptation and Innovations in the Indo-Gangetic Plains, South Asia*". This research was conducted by Southasia Institute of Advanced Studies (SIAS) of Nepal with local partner organization Centre for Research on Innovation and Science Policy (CRISP) India and Flood Hazard Research Centre, Bangladesh.

We would like to thank CGIAR Research Program on Climate Change, Agriculture and Food Security (CCAFS) for the funding support. During our research work, we were supported by different public and private institutions working in agricultural sectors. We are thankful to all of these institutions for providing valuable suggestions and information. We also express our deepest gratitude to all the farmers for their active participation and their effective support during our field work.

Author details

Dhanej Thapa[1]*, Yuba Raj Subedi[2] and Hemant Ojha[3,4]

*Address all correspondence to: dhanejthapa@hotmail.com

1 Nepal Development Research Institute (NDRI), Kathmandu, Nepal

2 ForestAction Nepal, Kathmandu, Nepal

3 University of New South Wales (UNSW), Sydney, Australia

4 Southasia Institute of Advance Studies, Kathmandu, Nepal

References

[1] Agrawal A. The role of local institutions in adaptation to climate change. International Forestry Research and Institutions Program (IFRI) Working Paper. 2008

[2] IPCC. Climate Change 2007: Synthesis Report. Contribution of Working Groups I, II, and III to the Fourth Assessment Report of the Intergovernmental Panel on Climate Change. In: Pacharuri RK, Reisinger A, editors. Intergovernmental Panel on Climate Change, Geneva, Switzerland; 2007. p. 104

[3] Rosenzweig C, Parry ML. Potential impact of climate change on world food supply. Nature. 1994;**367**:133-138

[4] Grace PR, Jain MC, Harrington L, Robertson GP. Long term sustainability of the tropical and subtropical rice-wheat system: An environmental perspective. In: Improving the Productivity and Sustainability of Rice-Wheat Systems: Issues and Impacts. ASA Special Publication 65. USA: ASSA-CSSA-SSSA; 2003

[5] Parry ML, Rosenzweig C, Iglesias A, Livermore M, Fischer G. Effects of climate change on global food production under SRES emissions and socio-economic scenarios. Global Environmental Change. 2004;**14**:53-67

[6] Warren R. Understanding the regional impacts of climate change. Research Report Prepared for the Stern Review. Norwich: Tyndall Centre Working Paper 90; 2006

[7] Kelkar U, Bhadwal S. South Asian regional study on climate change impacts and adaptation: Implications for human development. Human Development Report 2007/2008. UNDP; 2008

[8] Maplecroft. 2010. Available from: http://maplecroft.com [Accessed: 2013-01-12]

[9] Ministry of Environment (MOE). National Adaptation Program of Actions to Climate Change, Nepal. Kathmandu, Nepal: MOE, 2010

[10] World Bank. Nepal Country Environmental Analysis: Strengthening Institutions and Management Systems of Enhanced Environment Governance; The World Bank, Washington DC, USA; 2008

[11] CBS. Agricultural Census Survey, Nepal. Kathmandu, Nepal: Central Bureau of Statistics; 2011

[12] Shrestha AB, Wake CP, Mayewski PA, Dibb JE. Maximum temperature trends in the Himalaya and its vicinity: An analysis based on temperature records from Nepal for the period 1971-94. Journal of Climate. 1999;**12**:2775-2787

[13] Regmi HR. Effect of unusual weather on cereal crops production and household food security. The Journal of Agriculture and Environment. 2007;**8**:20-29

[14] Practical Action. Temporal and Spatial Variability of Climate Change over Nepal. Practical Action Nepal: Practical Action Nepal Office; 2009

[15] NCVST. Vulnerability through the Eyes of Vulnerable: Climate Change Induced Uncertainties and Nepal's Development Predicaments. Kathmandu: Institute for Social and Environmental Transition-Nepal (ISET-N, Kathmandu) and Institute for Social and Environmental Transition (ISET, Boulder, Colorado) for Nepal Climate Vulnerability Study Team (NCVST); 2009

[16] Asian Development Bank (ADB). Climate Change Assessment, Document Stage Draft for Review by Development Partners. Kathmandu, Nepal: ADB; 2009

[17] IDS. Economic Impact Assessment of Climate Change in Key Sectors in Nepal. IDS-Nepal: Kathmandu, Nepal; 2014

[18] ICIMOD. The Himalayan Climate and Water Atlas: Impact of Climate Change on Water Resources in Five of Asia's Major River Basins. Kathmandu, Nepal; December 2015

[19] Intergovernmental Panel on Climate Change (IPCC). Impacts, Adaptation and Vulnerability. Contribution of Working Group II to the Fourth Assessment Report of the Intergovernmental Panel on Climate Change; 2001

[20] Smit B, Skinner MW. Adaptation options in agriculture to climate change: A typology. Mitigation and Adaptation Strategies for Global Change. 2002;**7**:85-114

[21] Smit B, Burton I, Klein JT, Wandel J. An anatomy to adaptation to climate change and variability. Climatic Change. 2000;**45**(1):223-251

[22] Adger WN, Arnell NW, Tompkins EL, Successful adaptation to climate change across scales. Global Environmental Change. 2005;**15**(2):77-86. ISSN: 0959-3780. DOI: https://doi.org/10.1016/j.gloenvcha.2004.12.005

[23] Spielman DJ, Ekboir J, Davis K. The art and science of innovation systems inquiry: Applications to sub-Saharan African agriculture. Technology in Society. 2009;**31**(4):399-405

[24] Hall A, Clark N. What do complex adaptive systems look like and what are the implications for innovation policy? Journal of International Development. 2010;**22**:308-324

[25] Creswell JW. Research Design: Qualitative, Quantitative, and Mixed Methods Approach. London: Sage Publications; 2003

[26] Forch W, Kristjanson P, Thornton P, Kiplimo J. Initial sites in the CCAFS regions: Eastern Africa, West Africa and Indo-Gangetic Plains. In: CGIAR Research Program. Climate Change, Agriculture and Food Security; Copenhagen, Denmark 2011. pp.1-71

[27] Ministry of Agriculture and Development (MOAD). Statistical Information on Nepalese Agriculture. Kathmandu, Nepal: MOAD; 2015

[28] Thapa D. Climate risk and changing cropping pattern in rice-wheat system of Hattibangai VDC of Rupandehi District [MSc thesis]. Submitted to. Kathmandu, Nepal: School of Arts, Kathmandu University; 2012

[29] Manandhar S, Vogt DS, Perret SR, Kazama F. Adapting cropping systems to climate change in Nepal: A cross-regional study of farmers' perception and practices. Regional Environmental Change. 2011;**11**:335-348

[30] NWRP. Annual Progress Report. Bhairawa, Rupandehi, Nepal: National Wheat Research Program; 2011

[31] DADO. Annual Progress Report. Rupandehi, Nepal: District Agriculture Development Office; 2012

[32] Kristjanson P, Garlick C, Ochieng S, Förch W, Thornton PK. Global summary of baseline household survey results. In: CGIAR Research Program on Climate Change, Agriculture and Food Security (CCAFS). Copenhagen, Denmark; 2011

[33] Dahal KR, Poudyal CP, Adhikari P, Sharma S, Ghimre J. Effects of riverbed extraction on ground water resources in the vicinity of Tinau river, Rupandehi. Nepal Journal of Science and Technology. 2012;**13**(2):133-140

[34] Gauchan D, and Gumma MK. Effect of drought incidence on rice area, production and productivity in Nepal. Estimation of temporal data using remote sensing and socio-economic methods. A paper presented in IWMI National Conference on Water, Food Security and Climate Change, Nov 23-24, 2011; Kathmandu; 2009

[35] Amgain LP, Devkota NR, Timsina J, Singh B. Effect of climate change and CO_2 concentration on growth and yield of rice and wheat in Punjab: Simulations using CSM-CERES-rice and CSM-CERES-wheat models. Journal of Institute of Agriculture and Animal Science. 2006;**27**:103-110

[36] Aggarwal PK. Global climate change and Indian agriculture: Impacts, adaptation and mitigation. Indian Journal of Agricultural Sciences. 2008;**78**:911-919

[37] Adger WN. Social capital, collective action and adaptation to climate change. Economic Geography. 2003;**79**:387, 2003-2404

[38] Deniere A, Luong HV, editors. The Dynamics of Social Capital and Civic Engagement in Asia. London and New York: Recherche; 2012

[39] Ministry of Agriculture and Development (MOAD). Agriculture Development Strategy. Singhdurbar, Kathmandu, Nepal: MOAD; 2015

[40] Howden SM, Soussana JF, Tubiello FN, Chhetri N, Dunlop M, Meinke H. Adapting agriculture to climate change. Proceedings of the National Academy of Sciences of the United States of America. 2007;**104**:19691-19696

[41] Smit B, Wandel J. Adaptation, adaptive capacity and vulnerability. Global Environmental Change. 2006;**16**(3):282-292

[42] Tschakert P, Dietrich KA. Anticipatory learning for climate change adaptation and resilience. Ecology and Society. 2010;**15**(11). (online) http://www.ecologyandsociety.org/vol15/iss2/art11/

[43] Mezirow J. Transformative Dimensions of Adult Learning. San Francisco, California: Jossey-Bass; 1991

Chapter 3

Naturally Available Genetic Adaptation in Common Bean and Its Response to Climate Change

Andrés J. Cortés and Matthew W. Blair

Additional information is available at the end of the chapter

http://dx.doi.org/10.5772/intechopen.72380

Abstract

Warming is expected to lead to drier environments worldwide, especially in the tropics, and it is unclear how crops will react. Drought tolerance often varies at small spatial scales in natural ecosystems, where many of the wild relatives and landraces of the main crops have been collected. Through a series of examples, we will show that collections of wild relatives and landraces, many of those deposited at germplasm banks, may represent this desired source of variation, as they are genetically diverse and phenotypically variable. For instance, using a spectrum of genotyping and phenotyping approaches, we have studied the extent of genetic and phenotypic diversity for drought tolerance in wild and landraces of common bean (*Phaseolus vulgaris* L.) and compared it with the one available at cultivated varieties. Not surprisingly, most of the naturally available variation to cope with drought in the natural environments was lost through domestication and recent plant breeding. It is therefore imperative to exploit the reservoir of wild relatives and landraces to make crops more tolerant. Yet, it remains to be seen if the rate at which this naturally available variation can be incorporated into the cultivated varieties may keep pace with the rate of climate change.

Keywords: drought tolerance, environmental adaptation, genomic signatures of selection, agroecological models, divergent selection

1. Common bean: a model to explore the usefulness of wild relatives and landraces as a resource for the future

In the present chapter, we review the utility of genome-environment association approaches to infer the potential of wild accessions and landraces to make tropical crops more resistant to climate change, using the food crop common bean (*Phaseolus vulgaris* L.) as a model. Wild bean is thought to have diversified and adapted locally in South and Central

Figure 1. Geographic distribution of wild relatives (light gray) and landraces (dark gray) of common bean and its diversity in terms of seed size, colors, and patterns. Modified from Cortés *et al.* [12].

America from an original range in Central America [1, 2], after which domestication in the southern and northern ends of each region gave origin to Andean and Mesoamerican domesticates, respectively [3–7]. Both genepools followed somewhat parallel pathways of dissemination through the world, generating new secondary centers of diversity in Africa and Asia [8].

Common bean is a source of nutrients and protein for over 500 million people in Latin America and Africa, and more than 4.5 out of 23 million hectares are grown in zones where drought is severe, such as in northeastern Brazil, coastal Peru, the central and northern highlands of Mexico, and Eastern and Southern Africa [9, 10]. This situation may worsen as increased drought due climate change will reduce global crop production in >10% by 2050 [11]. Increasing drought tolerance in common bean varieties is therefore needed. Characterizing geo-referenced landraces and wild accessions of common bean at the genetic level (e.g., **Figure 1**) and quantifying SNP allelic associations with a bioclimatic-based drought index offer an efficient path to identify adaptive variation suitable to breed new drought-tolerant varieties.

In the following two sections, we first explain the theoretical bases behind genome-environment associations, as well as its caveats, (Section 2) and later we exemplify it with concrete cases that used geo-referenced landraces and wild accessions of common bean to infer naturally available adaptive variation (Section 3).

2. Strategies to infer adaptability of wild relatives and landraces to their natural habitats

Understanding the genomic signatures associated with environmental variation provides insights into how species adapt to their environment [13–15]. Recent genomic studies in wild populations have demonstrated that genome-environment associations, which are associations between single-nucleotide polymorphism (SNP) alleles and accessions' environment of origin, can indeed be used to identify adaptive loci and predict phenotypic variation. For instance, Turner and Bourne [16] predicted genetic adaptive variation to serpentine soils in *Arabidopsis lyrata*, Hancock and Brachi [17] identified climate-adaptive genetic loci among a set of geographically diverse *Arabidopsis thaliana*, Fischer and Rellstab [18] predicted adaptive variation to topo-climatic factors in *Arabidopsis halleri*, Pluess and Frank [19] predicted genetic local adaptation to climate at a regional scale in *Fagus sylvatica*, and Yeaman and Kathryn [20] detected convergent local adaptation in two distantly related species of conifers.

This genome-environment association approach has also been explored in some crop accessions as a prospection strategy of germplasm, alternative to traditional phenotyping. For example, Yoder and Stanton-Geddes [21] were able to capture adaptive variation to thermal tolerance, drought tolerance, and resistance to pathogens in *Medicago truncatula*; Lasky and Upadhyaya [22] predicted genotype-by-environment interactions to drought stress and aluminum toxicity in *Sorghum bicolor*; and Berthouly-Salazar and Thuillet [23] uncovered genomic regions involved in adaption to abiotic and biotic stress on two climate gradients in *Cenchrus americanus*.

Nonetheless, since genomic signatures associated with habitat heterogeneity can result from causes other than adaptation and selection [24, 25], for example, random genetic drift (**Figure 2**), and are also influenced by differences in ancestral variation and recombination in the genome [27–29], some further approaches need to be undertaken to clarify the truth nature of the divergent regions. For instance, the origin of habitat-associated variants from novel or standing genetic variation leads to distinctively different patterns of genomic divergence [30–32]. One approach that can help to distinguish these underlying causes of divergence is comparing summary statistics (i.e., Tajima's D) from different genomic sections because demographic processes usually leave genome-wide signatures while selection tends to imprint more localized regions [33]. Specifically, habitat-mediated purifying selection is associated with localized low values of nucleotide diversity (π) [34] and Tajima's D [35] and high scores of the Watterson's theta (θ) estimator [36] because only low-frequency polymorphisms can avoid being eliminated by widespread directional selection. Although recent population bottlenecks tend to achieve the same reduction in nucleotide variation, this pattern is expected at a more genome-wide level. Similarly, local adaptation tends to homogenize haplotypes within the same niche, fix polymorphisms in different populations, and eliminate low-frequency polymorphism. Consequently, few haplotypes with high frequency are retained, corresponding to high values of nucleotide diversity (π) and Tajima's D and low scores of the

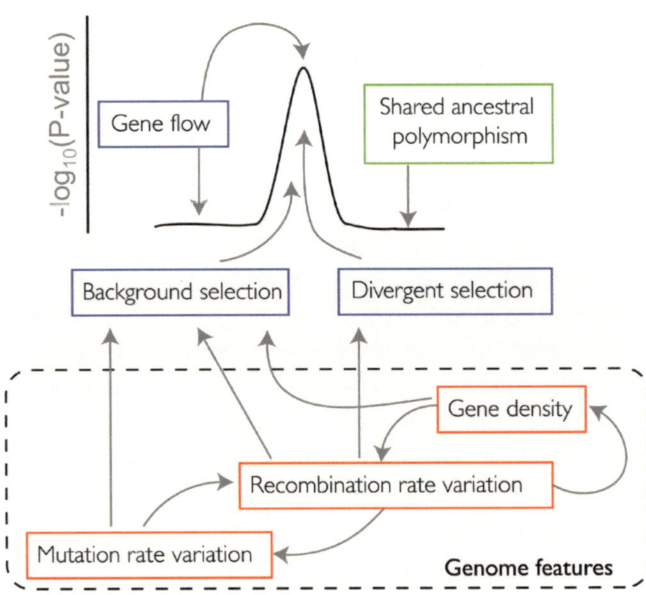

Figure 2. Multiple causes explain genome-environment associations. External processes, such as divergent selection, which is the main focus when assessing adaptation in wild relatives and landraces of crops, is only one of many possible causes. At the same time, the genomic background may be homogenized by gene flow [26]. Similarly, background selection and genomic features in regions of reduced recombination rate and shared ancestral polymorphism (more prone to genetic drift due to their reduced effective population size) could induce hotspots of spurious genome-environment associations. Therefore, besides external processes driven by natural selection, both inherent properties of the genome and the demographic and evolutionary history of the crop influence the extent of the genome-environment associations. Modified from Ravinet, Faria [64].

Watterson's theta (θ) estimator [33]. Although independent domestication events, extensive population structure, and population expansions after bottlenecks can produce the same patterns, these demographic processes also imprint genomes at a more genome-wide level.

In the following two subsections, we explain how to implement genome-environment associations in order to infer adaptability of wild relatives and landraces to their natural habitats (Section 2.1) and discuss ways to account for causes, other than adaptation and selection that may be shaping the genomic landscape of signatures associated with habitat heterogeneity (Section 2.2).

2.1. Using genome-environment association scans to identify loci associated with bioclimatic-based indexes

First of all, in order to account for possible demographic effects, subpopulation structure must be determined in geo-referenced landraces and wild accessions using principal coordinate analysis (PCoA) implemented in the software Trait Analysis by aSSociation, Evolution and Linkage, Tassel v.5 [37]. The same dataset and software can be used to perform association analyses between the SNP markers and bioclimatic-based indexes (e.g., [12, 38, 39]).

As a rule of thumb, a total of 10 generalized (GLM) and mixed linear models (MLM) should be compared [40]. Within each model family, five models are usually built as follows: (1) model with the genepool identity and the first two PCoA axes scores as covariates; (2) models with the within-genepool subpopulation identity (e.g., [41]) and the first two PCoA axes scores as covariates; (3) model with the first two PCoA axes scores as covariates; (4) model with the within-genepool subpopulation identity (e.g., [41]), as covariate; and (5) model with the genepool identity as covariate. All five MLMs usually use a centered IBS kinship matrix as a random effect to control for genomic background implementing the EMMA and P3D algorithms to reduce computing time [42]. QQ-plots of the P-values should be inspected to assess whether excessive numbers of false positives are generated and choose in this way the optimum model. Significant associations are determined using strict Bonferroni corrections of P-values at alpha = 0.001, leading, for example, to a significance threshold of 4.4×10^{-8} in a usual dataset of ca. 23,000 SNP markers (0.001 divided by the number of markers) or $-\log_{10}(4.4 \times 10^{-8}) = 7.36$. The construction of customized PCoA and Manhattan diagrams can be carried out with the software R v.3.3.1 (R Core Team).

Finally, candidate genes for habitat adaptation can be identified within the 1000 bp sections flanking each SNP marker that is associated with a bioclimatic-based index by using the corresponding reference genome (e.g., [5]) and the PhytoMine and BioMart tools in Phytozome v.12 (phytozome.jgi.doe.gov).

2.2. Accounting for genomic constrains by inspecting genome-wide patterns of variation

In order to identify causes other than adaptation and selection that may be shaping the genomic landscape of signatures associated with habitat heterogeneity (i.e., genomic constrains and genetic drift), sliding window approaches (e.g., window size = 1×10^6 bps, step size = 200 kb) can be implemented to describe patterns of variation and overall divergence across the genome. For instance, SNP density, nucleotide diversity as measured by π [34], Watterson's theta (θ) estimator [36], and Tajima's D [35] can be computed using the software Tassel v.5 [37] and customized R scripts. Results of all windowed analyses are usually plotted against window midpoints in millions of base pairs (Mb) in the software R v.3.3.1 (R Core Team). The centromeres can be marked to visualize the extent of the centromeric repeats and its correlation with overall patterns of diversity and divergence.

It is advisable to calculate bootstrap-based means and 95% confidence intervals around the mean for some summary statistics (i.e., SNP density, π, θ, and Tajima's D) when computed in sliding windows that contained or did not contain at least one marker that was associated with a bioclimatic-based index. For this, each summary statistic of windows containing and not containing associated SNPs should be randomly resampled with replacement (bootstrapping) across windows within grouping factor (associated vs. no associated). The overall mean is then stored for each grouping factor. This step should iterated at least 1000 times using customized R scripts. Bootstrapping must be performed independently for each summary statistic in order to eliminate correlations among these.

3. The adaptive potential of wild relatives and landraces in common bean

In common bean, ecological gradients related with drought stress are associated with divergent selection at the genetic level, after accounting for genepool and subpopulation structure. This divergent selective pressure might be a consequence of local-level rainfall patterns. Specifically, in tropical environments near the equator with bimodal rainfall, a mid-season dry period occurs that can last to 2–4 weeks. In contrast in the subtropics, a dry period of three or more months can occur. In response to this mid-cycle drought of the subtropics, *P. vulgaris* enters a survival mode of slow growth and reduced physiological activity until rainfall resumes and flowering occurs [43]. Beans growing in wetter conditions on the other hand are less frequently subjected to these environmental pressures and have a fitness advantage to mature in a shorter length of time. Given these ecological differences, and consistent with genomic signatures of divergent selection, the reaction typically associated with drought tolerance although favorable under dry conditions seems detrimental under more humid conditions. The awareness about this trade-off may aid the breeding of new drought-tolerant varieties specifically adapted to unique microenvironments (e.g., [44]) and local regions rather than varieties eventually obsolete, originally intended for a wider range of environments.

In the next two subsections, we summarize the concrete evidence supporting these statements (Section 3.1) and explain how we can discard other fortuitous causes that may also explain the same pattern (Section 3.2), based on the approaches that we introduced in the previous section (Section 2).

3.1. The signatures of adaptation in common bean are widespread throughout the genome

SNP markers are good at recovering the well-described Andean and Mesoamerican genepool structure and the five within-genepool subpopulations observed in wild common bean [41].

Because of this, in a previous research from us with more than 22,000 SNP markers, QQ-plots from the association analyses between those SNP markers and a bioclimatic-based drought index [12] indicated that GLM analyses likely had excessive rates of false positives, whereas MLM models controlling for population structure and using a kinship matrix reduced more effectively the false-positive rate.

In that particular case, the MLM model with the first two PCoA axes scores used as covariates was the best at controlling for false positives. This model yielded a total of 115 SNP markers associated with the bioclimatic-based drought index at a Bonferroni-corrected significance threshold of 7.36 $-\log_{10}$(P-value). These markers explained on average 51.3% ± 0.4 of the variation in the bioclimatic-based drought index. The 115 SNPs were clustered in 90 different regions, defined as overlapping 1000 bp sections that flanked associated markers (**Figure 3**). Associated SNPs and regions were widespread in all 11 common bean chromosomes.

Following the previous example, chromosomes Pv3 and Pv8 had the highest number of associated SNPs with 21 and 32 SNPs clustered in 16 and 21 different regions, respectively. Chromosomes Pv1, Pv2, Pv4, Pv5, Pv6, and Pv9 contained an intermediate number of associated SNPs with 11, 6, 11, 7, 12, and 9 SNPs clustered in 11, 6, 8, 6, 8, and 9 different regions, respectively. Chromosomes Pv7, Pv10, and Pv11 had the fewest number of associated SNPs with 3, 2, and 1 SNPs clustered in 3, 1, and 1 different regions, respectively. Chromosome Pv8 had more regions with at least two associated SNPs than any other chromosome, and these regions had more associated SNPs than in any other chromosome for a total of five regions with an average number of associated SNPs of 3.2. The single region that contained more associated SNPs was also situated in chromosome Pv8 with six SNPs explaining on average 51.1% ± 0.3 of the variation in the bioclimatic-based drought index. After chromosome Pv8, Pv3 was also outstanding having four regions (with at least two associated SNPs) with an average number of associated SNPs of 2.5. Therefore, a total of 75 regions, comprising 99 SNP markers associated with the bioclimatic-based drought index, contained at least 1 gene, for a total of 77 genes. Most genes were in chromosomes Pv1, Pv3, and Pv8 with 11, 14, and 16 genes. Only two regions, at chromosomes Pv1 and Pv8 and containing a total of seven different SNPs, spanned two or more genes. The one in Pv8 was the region with more associated SNPs (six in total). One of the two genes in this region encoded an Ankyrin repeat-containing protein, which was associated with osmotic regulation via the assembly of cation channels in the membranes [45]. Among other identified candidate genes, there was a phototropic-responsive NPH3 gene [46] in Pv3.

3.2. Rampant divergent selection: interpreting genomic signatures of adaptation in common bean beyond genomic constrains

As a follow-up of the previous example, associated genomic windows were enriched for SNP density and positive Tajima's D scores. This conclusion was achieved after implementing a sliding window analysis to explore the patterns of genome-wide diversity (**Figure 3**). Marker density decayed drastically toward the centromeres. This decay in diversity proportional to the decay in the rate of recombination was first described in *D. melanogaster* and has been confirmed in many organisms since then. The correlation was initially understood as an effect of genetic hitchhiking, but background selection has been increasingly appreciated as a contributing factor [28], perhaps in many cases the dominating one.

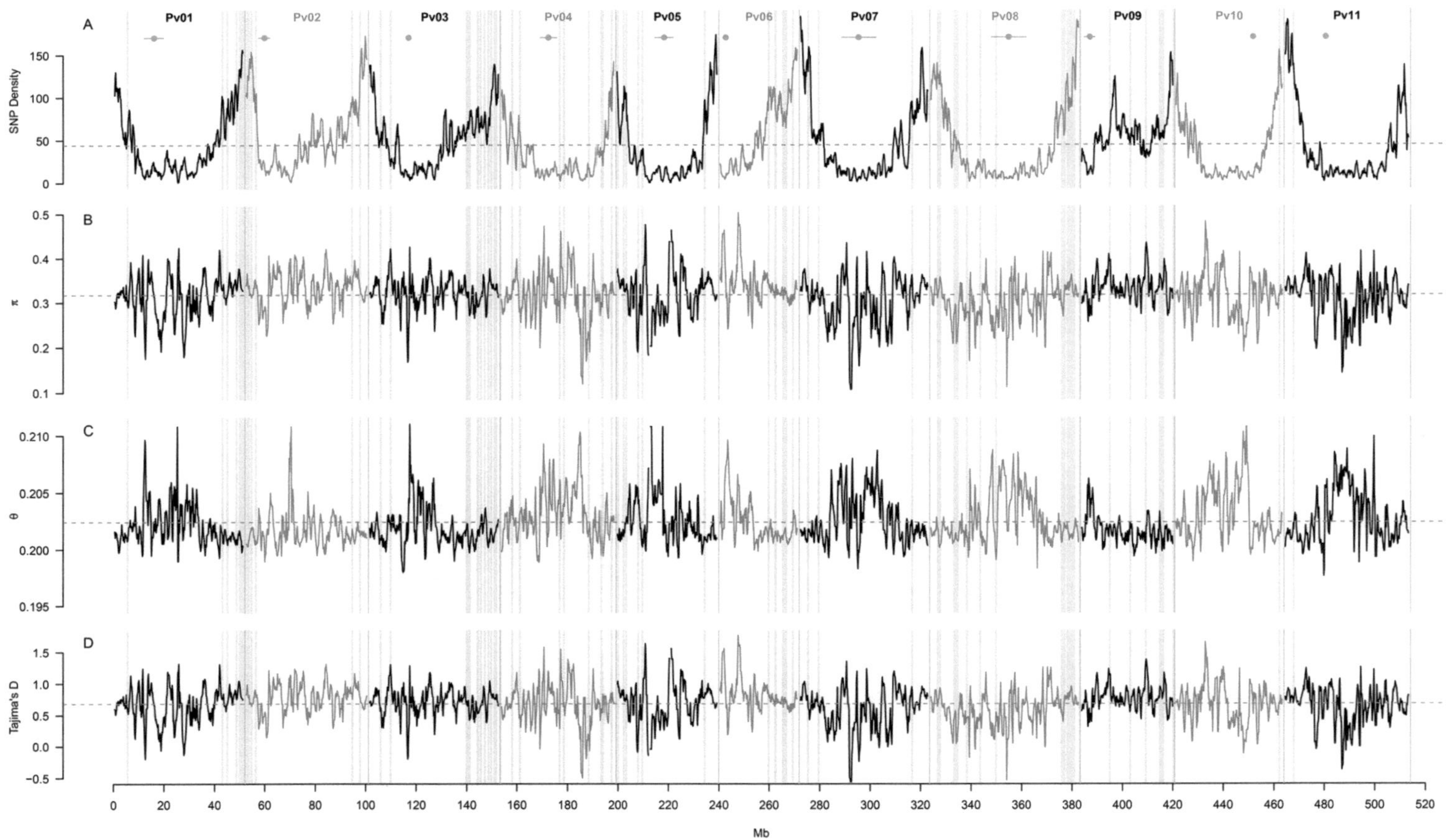

Figure 3. Patterns of genome-wide diversity in common bean accessions based on SNP markers. A sliding window analysis was used to compute **(A)** SNP density, **(B)** nucleotide diversity as measured by π, **(C)** Watterson's theta estimator (θ), and **(D)** Tajima's D. Results of all windowed analyses are plotted against window midpoints in millions of base pairs (mb). Black and gray colors highlight different common bean (pv) chromosomes. Gray-dashed horizontal lines indicate genome-wide averages. Gray vertical boxes indicate the 1000 bp flanking region of each marker that was associated with the bioclimatic-based drought index. Horizontal gray lines with a central filled gray dot at the top of the figure mark the centromeres according to Schmutz, McClean [5].

Average marker density was 44 SNPs per million base pairs (95% CI, 4–143). Average nucleotide diversity as measured by π was 0.3 per million base pairs (95% CI, 0.2–0.4). Average Watterson's theta (θ) was 0.20 per million base pairs (95% CI, 0.19–0.21). Average Tajima's D was 0.68 per million base pairs (95% CI, 0.05–1.22). These very same statistics were compared between 1 Mb sliding windows that contained (associated) or did not contain (no associated) at least one marker that was associated with the bioclimatic-based drought index. Genomic windows containing at least one associated SNP had overall higher SNP density (79 ± 6 vs. 39 ± 2), lower values for Watterson's theta (θ) scores (0.2016 ± 0.0001 vs. 0.2026 ± 0001), and more positive Tajima's D scores (0.71 ± 0.02 vs. 0.678 ± 0.009) than windows without associated markers. Nucleotide diversity, as measured by π, was slightly elevated in associated windows when compared with no associated windows (0.322 ± 0.006 vs. 0.317 ± 0.003).

Selective process, such as purifying selection and local adaptation (divergent selection), differentially imprints regions within the same genome, causing a heterogeneous departure of genetic variation from the neutral expectations and from the background trend [28]. Divergent selection tends to homogenize haplotypes within the same niche, fix polymorphisms in different populations, and eliminate low-frequency polymorphism. Consequently, few haplotypes with high frequency are retained, corresponding to high values of nucleotide diversity and Tajima's D and low scores of the Watterson's theta (θ) estimator [33]. We have identified these signatures in the various genomic regions associated with a bioclimatic-based drought index. Therefore, it is unlikely that independent domestication events, extensive population structure, and population expansions after bottlenecks are responsible for these patterns because the mixed linear model that we used to identify the genome-environment associations accounted for population structure, while demographic processes would leave genome-wide signatures in both, associated and no associated windows.

4. Conclusions

Wild accessions and landraces of common bean occupy more geographical regions with extreme ecologies [2] and extensive drought stress [12] than cultivated accessions. Those regions include the arid areas of Peru, Bolivia and Argentina, and the valleys of northwest Mexico. Hence, a broad habitat distribution for wild common bean has exposed these genotypes to both dry and wetter conditions, while cultivated common bean has a narrower distribution and is traditionally considered susceptible to drought. These differences in the ecologies of wild and cultivated common bean have been associated with higher genetic diversity in the former group when surveying candidate genes for drought tolerance such as the ASR [47], DREB [48], and ERECTA [49] gene families, once population structure [41] and the background distribution of genetic diversity have been accounted for.

Also, as identified through the genome-environment association approach that was illustrated in this chapter, there are notorious differences between the adaptations of wild accessions and landraces found in arid and more humid environments, in congruence with natural divergent selection acting for thousands of years. Several of these differences might be valuable for plant breeding. Therefore, we reinforce, as was envisioned by Acosta and Kelly [50], that wild accessions and landraces of common bean be taken into account to exploit naturally available divergent variation for drought tolerance. We envision that this lesson from common bean will inspire the exploitation of wild relatives and landraces of other crops to face the threats imposed by current climate change.

5. Prospects

This chapter ultimately illustrates that genomic signatures of environmental adaptation (e.g., [51]) are useful for germplasm characterization, potentially enhancing future marker-assisted selection and crop improvement. We envision that genome-environment association studies coupled with estimates of genome-wide diversity will become more common in the oncoming years. These types of studies will likely go beyond estimates of drought tolerance, as exemplified here, to also include estimates regarding frost stress (i.e., [52–54]), nutrient limitation [55, 56], as well as other threats imposed by climate change [57, 58] in different types of ecosystems (e.g., [59]) and screened by a variety/wide range of genotyping techniques [60–63]. Genomic selection models [64] could also incorporate at some point environmental variables in order to improve the prediction of phenotypic variation and the estimation of the genotype-by-environment interactions [65] in the light of linkage disequilibrium (LD) [66] and various stochastic models [67, 68].

Acknowledgements

We are grateful with Daniel G. Debouck and the Genetic Resources Unit at the International Center for Tropical Agriculture for donating the seeds and sharing the geographic coordinates of the plant material that was refer to in this study. This research was supported by the Lundell and Tullberg grants to AJC. The Geneco mobility fund from Lund University is acknowledged for making possible the synergistic meeting between AJC and MWB in the spring of 2015 at Nashville, TN.

Author details

Andrés J. Cortés[1,2]* and Matthew W. Blair[3]

*Address all correspondence to: acortes@corpoica.org.co

1 Department of Biological and Environmental Sciences, University of Gothenburg, Gothenburg, Sweden

2 Colombian Corporation for Agricultural Research (Corpoica), Rionegro, Colombia

3 Department of Agricultural & Environmental Science, Tennessee State University, TN, Nashville, USA

References

[1] Bitocchi E et al. Mesoamerican origin of the common bean (*Phaseolus vulgaris L.*) is revealed by sequence data. Proceedings of the National Academy of Sciences. 2012;**109**(14):E788-E796

[2] Rodriguez M et al. Landscape genetics, adaptive diversity and population structure in *Phaseolus vulgaris*. The New Phytologist. 2016;**209**(4):1781-1794

[3] Gepts P, Debouck D. Origin, domestication and evolution of the common bean (*Phaseolus vulgaris L.*). In: Van Shoonhoven A, Voysest O, editors. Common beans: Research for crop improvement. Wallingford: Commonwealth Agricultural Bureau; 1991. pp. 7-53

[4] Kwak M, Gepts P. Structure of genetic diversity in the two major gene pools of common bean (*Phaseolus vulgaris* L., *Fabaceae*). Theoretical and Applied Genetics. 2009;**118**(5):979-992

[5] Schmutz J et al. A reference genome for common bean and genome-wide analysis of dual domestications. Nature Genetics. 2014;**46**(7):707-713

[6] Bitocchi E et al. Molecular analysis of the parallel domestication of the common bean (*Phaseolus vulgaris*) in mesoamerica and the Andes. New Phytologist. 2013;**197**(1):300-313

[7] Cortés AJ. On the origin of the common bean (*Phaseolus Vulgaris L.*). American Journal of Plant Sciences. 2013;**4**(10):1998-2000

[8] Asfaw A, Blair MW, Almekinders C. Genetic diversity and population structure of common bean (*Phaseolus vulgaris L.*) landraces from the east african highlands. Theoretical and Applied Genetics. 2009;**120**(1-12)

[9] Broughton WJ et al. Beans (phaseolus spp.) - model food legumes. Plant and Soil. 2003;**252**(1):55-128

[10] Darkwa K et al. Evaluation of common bean (*Phaseolus vulgaris L.*) genotypes for drought stress adaptation in Ethiopia. The Crop Journal. 2016;**4**(5):367-376

[11] Tai APK, Martin MV, Heald CL. Threat to future global food security from climate change and ozone air pollution. Nature Climate Change. 2014;**4**:817-821

[12] Cortés AJ et al. Drought tolerance in wild plant populations: The case of common beans (*Phaseolus Vulgaris L.*). PLoS One. 2013;**8**(5):e62898

[13] Hoffmann AA, Sgro CM. Climate change and evolutionary adaptation. Nature. 2011;**470**(7335):479-485

[14] Stinchcombe JR, Hoekstra HE. Combining population genomics and quantitative genetics: Finding the genes underlying ecologically important traits. Heredity. 2008;**100**:158-170

[15] Savolainen O, Lascoux M, Merilä J. Ecological genomics of local adaptation. Nature Reviews Genetics. 2013;**14**(11):807-820

[16] Turner TL et al. Population resequencing reveals local adaptation of *Arabidopsis lyrata* to serpentine soils. Nature Genetics. 2010;**42**(3):260-263

[17] Hancock AM et al. Adaptation to climate across the *Arabidopsis thaliana* genome. Science. 2011;**334**(6052):83-86

[18] Fischer MC et al. Population genomic footprints of selection and associations with climate in natural populations of *Arabidopsis halleri* from the alps. Molecular Ecology. 2013;**22**(22):5594-5607

[19] Pluess AR et al. Genome-environment association study suggests local adaptation to climate at the regional scale in *Fagus sylvatica*. The New Phytologist. 2016;**210**(2):589-601

[20] Yeaman S et al. Convergent local adaptation to climate in distantly related conifers. Science. 2016;**353**(6306):1431-1433

[21] Yoder JB et al. Genomic signature of adaptation to climate in *Medicago truncatula*. Genetics. 2014;**196**(4):1263-1275

[22] Lasky JR et al. Genome-environment associations in sorghum landraces predict adaptive traits. Science Advances. 2015;**1**(6)

[23] Berthouly-Salazar C et al. Genome scan reveals selection acting on genes linked to stress response in wild pearl millet. Molecular Ecology. 2016;**25**(21):5500-5512

[24] Nei M. The neutral theory of molecular evolution in the genomic era. Annual Review of Genomics and Human Genetics. 2010;**11**:265-289

[25] Cortés AJ et al. Genome-wide patterns of microhabitat-driven divergence in the alpine dwarf shrub *Salix herbacea L.* In: On The Big Challenges of a Small Shrub: Ecological Genetics of *Salix herbacea L.* Uppsala: Acta Universitatis Upsaliensis; 2015

[26] Ravinet M et al. Interpreting the genomic landscape of speciation: A road map for finding barriers to gene flow. Journal of Evolutionary Biology. 2017;**30**(8):1450-1477

[27] Strasburg JL et al. What can patterns of differentiation across plant genomes tell us about adaptation and speciation? Philosophical Transactions of the Royal Society B: Biological Sciences. 2011;**367**(1587):364-373

[28] Ellegren H, Galtier N. Determinants of genetic diversity. Nature Reviews. Genetics. 2016;**17**(7):422-433

[29] Wolf JB, Ellegren H. Making sense of genomic islands of differentiation in light of speciation. Nature Reviews. Genetics. 2017;**18**:87-100

[30] Barrett RD, Schluter D. Adaptation from standing genetic variation. Trends in Ecology & Evolution. 2008;**23**(1):38-44

[31] Pritchard JK, Pickrell JK, Coop G. The genetics of human adaptation: Hard sweeps, soft sweeps, and polygenic adaptation. Current Biology. 2010;**20**(4):R208-R215

[32] Hermisson J, Pennings PS. Soft sweeps: Molecular population genetics of adaptation from standing genetic variation. Genetics. 2005;**169**(4):2335-2352

[33] Wakeley J. Coalescent Theory: An Introduction. Cambridge: Harvard University; 2008. p. 41

[34] Nei M. Molecular evolutionary genetics. New York: 1987. p. 50

[35] Tajima F. *Statistical method for testing the neutral mutation hypothesis by DNA polymorphism*. Genetics. 1989;**123**(3):585-595

[36] Watterson GA. Number of segregating sites in genetic models without recombination. Theoretical Population Biology. 1975;**7**:256-276

[37] Bradbury PJ et al. TASSEL: Software for association mapping of complex traits in diverse samples. Bioinformatics. 2007;**23**(19):2633-2635

[38] Cortés AJ. Environmental Heterogeneity at a Fine Scale: Ecological and Genetic Implications in a Changing World. Saarbrücken, Germany: LAP Lambert Academic Publishing; 2016. p. 60

[39] Cortés AJ. Local Scale Genetic Diversity and its Role in Coping with Changing Climate. In: Bitz L, editor. Genetic Diversity. InTech: Rijeka; 2017 140 pp

[40] Cortés AJ et al. Maintenance of female-bias in a polygenic sex determination system is consistent with genomic conflict. In: On The Big Challenges of a Small Shrub: Ecological Genetics of *Salix herbacea L*. Acta Universitatis Upsaliensis: Uppsala; 2015. p. 4

[41] Blair MW, Soler A, Cortés AJ. Diversification and population structure in common beans (*Phaseolus vulgaris L.*). PLoS One. 2012;**7**(11):e49488

[42] Zhang Z et al. Mixed linear model approach adapted for genome-wide association studies. Nature Genetics. 2010;**42**(4):355-360

[43] Beebe S et al. Selection for drought resistance in common bean also improves yield in phosphorus limited and favorable environments. Crop Science. 2008;**48**:582-592

[44] Cortés AJ et al. Small-scale patterns in snowmelt timing affect gene flow and the distribution of genetic diversity in the alpine dwarf shrub *Salix herbacea*. Heredity. 2014;**113**:233-239

[45] Voronin DA, Kiseleva EV. Functional role of proteins containing ankyrin repeats. Cell and Tissue Biology. 2008;**2**(1):1-12

[46] Pedmale UV, Liscum E. Regulation of phototropic signaling in arabidopsis via phosphorylation state changes in the phototropin 1-interacting protein NPH3. Journal of Biological Chemistry. 2007;**282**:19992-20001

[47] Cortés AJ et al. Molecular ecology and selection in the drought-related asr gene polymorphisms in wild and cultivated common bean (*Phaseolus Vulgaris L.*). BMC Genetics. 2012;**13**:58

[48] Cortés AJ et al. Nucleotide diversity patterns at the drought-related DREB2 encoding genes in wild and cultivated common bean (*Phaseolus Vulgaris L.*). Theoretical and Applied Genetics. 2012;**125**(5):1069-1085

[49] Blair MW, Cortés AJ, This D. Identification of an ERECTA gene and its drought adaptation associations with wild and cultivated common bean. Plant Science. 2016;**242**:250-259

[50] Acosta JA, Kelly JD, Gepts P. Prebreeding in common bean and use of genetic diversity from wild germplasm. Crop Science. 2007;**47**(S3):S44-S59

[51] Cortés AJ. On the Big Challenges of a Small Shrub: Ecological Genetics of *Salix herbacea L.* Uppsala: Acta Universitatis Upsaliensis; 2015. p. 14

[52] Wheeler JA et al. The snow and the willows: accelerated spring snowmelt reduces performance in the low-lying alpine shrub *Salix herbacea*. Journal of Ecology. 2016;**104**(4): 1041-1050

[53] Wheeler JA et al. Increased spring freezing vulnerability for alpine shrubs under early snowmelt. Oecologia. 2014;**175**(1):219-229

[54] Wheeler JA et al. With a little help from my friends: Community facilitation increases performance in the dwarf shrub *Salix herbacea*. Basic and Applied Ecology. 2015;**16**:202-209

[55] Little CJ et al. Small-scale drivers: The importance of nutrient availability and snowmelt timing on performance of the alpine shrub *Salix herbacea*. Oecologia. 2016;**180**(4):1015-1024

[56] Sedlacek J et al. What role do plant-soil interactions play in the habitat suitability and potential range expansion of the alpine dwarf shrub *Salix herbacea*? Basic and Applied Ecology. 2014;**15**(4):305-315

[57] Sedlacek J et al. Evolutionary potential in the alpine: Trait heritabilities and performance variation of the dwarf willow *Salix herbacea* from different elevations and microhabitats. Ecology and Evolution. 2016;**6**(12):3940-3952

[58] Sedlacek J et al. The response of the alpine dwarf shrub *Salix herbacea* to altered snowmelt timing: Lessons from a multi-site transplant experiment. PLoS One. 2015;**10**(4):e0122395

[59] Madriñán S, Cortés AJ, Richardson JE. Páramo is the world's fastest evolving and coolest biodiversity hotspot. Frontiers in Genetics. 2013;**4**(192)

[60] Blair MW et al. A high-throughput SNP marker system for parental polymorphism screening, and diversity analysis in common bean (*Phaseolus Vulgaris L.*). Theoretical and Applied Genetics. 2013;**126**(2):535-548

[61] Kelleher CT et al. SNP discovery, gene diversity, and linkage disequilibrium in wild populations of *Populus tremuloides*. Tree Genetics & Genomes. 2012;**8**(4):821-829

[62] Galeano CH et al. Gene-based single nucleotide polymorphism markers for genetic and association mapping in common bean. BMC Genetics. 2012;**13**(1):48

[63] Cortés AJ, Chavarro MC, Blair MW. SNP marker diversity in common bean (*Phaseolus Vulgaris L.*). Theoretical and Applied Genetics. 2011;**123**(5):827-845

[64] Desta ZA, Ortiz R. Genomic selection: Genome-wide prediction in plant improvement. Trends in Plant Science. 2014;**19**(9):592-601

[65] Cortés AJ, Blair MW. Genotyping by sequencing and genome – Environment associations in wild common bean predict widespread divergent adaptation to drought. Frontiers in Plant Science. 2018;**9**(128)

[66] Blair MW, Cortés AJ, Farmer A, Assefa T, Penmetsa RV, Cook D, et al. Uneven recombination rate and linkage disequilibrium across a reference SNP map for common bean. PLOS ONE. 2018; DOI: pone.0189597

[67] Cortés AJ. Prevalence in MSM is enhanced by role versatility. In: Mazari A, editor. Big Data Analytics in HIV/AIDS Research. Hershey: IGI Global; 2018

[68] Cortés AJ. On how role versatility boosts an STI. Journal of Theoretical Biology. 2017;**440**: 66-69

Chapter 4

Mitigation of the Negative Impact of Warming on the Coffee Crop: The Role of Increased Air [CO_2] and Management Strategies

Danielly Dubberstein, Weverton P. Rodrigues, José N. Semedo, Ana P. Rodrigues, Isabel P. Pais, António E. Leitão, Fábio L. Partelli, Eliemar Campostrini, Fernando Reboredo, Paula Scotti-Campos, Fernando C. Lidon, Ana I. Ribeiro-Barros, Fábio M. DaMatta and José C. Ramalho

Additional information is available at the end of the chapter

http://dx.doi.org/10.5772/intechopen.72374

Abstract

Crop sustainability can be threatened by new environmental challenges regarding predicted climate changes and global warming. Therefore, the study of real biological impacts of future environmental conditions (e.g., increased air [CO_2], supra-optimal temperature and water scarcity) on crop plants, as well as the re-evaluation of management procedures and strategies, must be undertaken in order to improve crop adaptation and promote mitigation of negative environmental impacts, thus affording crop resilience. Coffee is a tropical crop that is grown in more than 80 countries, making it one of the world's most traded agricultural products, while involving millions of people worldwide in the whole chain of value. It has been argued that this crop will be highly affected by climate changes, resulting in decreases in both suitable areas for cultivation and productivity, as well as impaired beverage quality in the near future. Here, we report recent findings regarding coffee species exposure to combined supra-optimal air temperatures and enhanced air [CO_2], and impacts of drought stress on the crop. Ultimately, we discuss key strategies to improve coffee performance in the context of new environmental scenarios. The recent findings clearly show that high [CO_2] has a positive impact on coffee plants, increasing their tolerance to high temperatures. This has been related to a better plant vigor, to the triggering of protective mechanisms, and to a higher functional status of the photosynthetic machinery. Even so, coffee plant is expected to suffer from water scarcity in a changing world. Therefore, discussion is focused on some important management strategies (e.g., shade systems, crop management and soil covering and terracing), which can be implemented to improve coffee performance and sustain coffee production in a continually changing environment.

Keywords: coffee crop sustainability, climate changes, mitigation, heat stress, drought

1. Introduction

Global emissions of the main greenhouses gases in the Earth's atmosphere raised in the mid-eighteenth century during the industrial revolution associated with the use of fossil fuels. Since then, the CO_2 concentration [CO_2] has increased from 280 to 400 μL CO_2 L^{-1} in 2014, and it is expected to rise to values between *ca.* 730 and 1020 μL CO_2 L^{-1} by 2100 [1]. Agricultural activity has also directly contributed to this process, being responsible for 1/3 of the CO_2 emissions but also with additional N_2O and CH_4 production, intensified mainly by inadequate management of crops and pastures [2], especially in low- and middle-income countries with predominating family farming [3].

Increased greenhouse gas emissions are expected to cause a temperature rise between 0.3 and 4.8°C by 2100, depending on future emissions and adequate measures to strongly limit them. Altered temperature may further promote extreme weather events, alter intra- and inter-annual precipitation patterns with long periods of drought and/or heavy rainfalls, partial melting of glacial ice, and consequently rising of the sea level [1]. Climate changes, particularly global warming, has a severe impact on the Earth's ecosystem and pose serious threats to agricultural sustainability [4–6], which is one of the human activities most vulnerable to climatic variation, since plants require optimal growing conditions to produce desired quantity and quality products [7, 8]. On the other hand, global demand for food is increasing as it is linked to the rapidly growing populations, which together with climate constraints, may compromise world food security [9]. In addition, increase in [CO_2] can affect the fundamental plant processes, such as photosynthesis and respiration, and, therefore, growth is also anticipated to be affected accordingly [10–12].

With regard to the coffee crop, it is known that plant growth, development, and productivity, as well as bean quality, are highly sensitive to climatic conditions [3, 13–16]. Accordingly, recent modeling studies have predicted important reductions of suitable areas for coffee cultivation in several producing regions [7, 16–19], with severe productivity losses in Mexico [20, 21], Nicaragua [3] and Tanzania [22], and extinction of wild populations of *C. arabica* in Ethiopia [23]. Although world coffee production has increased significantly in recent decades [24], studies state that climate change has caused substantial production losses [18], associated with periods of extreme droughts combined with supra-optimal temperatures [22, 25, 26], reducing coffee yields and bean quality as well as increasing the incidence of pests and diseases [16, 27]. In fact, it is believed that the recognized present climate changes have already caused yield losses in several coffee-producing countries, including Brazil, Ethiopia and Tanzania [22, 28, 29].

The negative estimates of future impacts on coffee crop were based on modeling approaches mostly focusing on increased air temperatures. However, these studies have only taken into account the current cultivars [30], and did not consider the considerable ability of some genotypes to endure various environment constraints, through metabolic adjustments and morphological and anatomical changes. Additionally, it was recently reported that coffee plants can respond positively to increased air [CO_2] [31–33], improving plant physiological and metabolic performance, and mitigating warming impacts [11, 33–35]. Such beneficial effect could even overcome these impacts, allowing some yield increase under adequate water availability to the crop [36], particularly at higher altitudes [37]. Nevertheless, given that coffee is one of the most important agronomic products, and that possible implications of ongoing climate changes may affect the sustainability of this crop in many actual areas, with potential dramatic economic, social and environmental implications, there is an urgent need for improving our knowledge regarding the plant performance under a wide range of environmental conditions. It is also equally important to identify adequate mitigation and adaptation strategies to be implemented, such as shading system crop management and soil cover and terracing, together with breeding new cultivars, in order to alleviate the impacts of climate changes on coffee plants.

Studies dealing with water stress in coffee species and genotypes have provided a detailed picture of biological mechanisms involved in drought tolerance [38–48], whereas recent works also showed that some genotypes of both *C. arabica* and *C. canephora* can endure temperatures much higher than what was traditionally accepted [11, 35]. As referred, plant resilience can even be improved under the exposure to high atmospheric [CO_2] [11, 34–37, 49]. In this context, the objective of this review chapter is to report the recent findings regarding the coffee plant responses to the single and combined exposure to atmosphere-supra-optimal temperatures and [CO_2], as well as to drought stress, together with the envisagement of some important crop management strategies (e.g., intercropping/shade systems, soil covering and terracing), which can be implemented to improve coffee performance and to mitigate the impact of environmental constraints, aiming at sustaining coffee production in a permanent changing environment.

2. General aspects of production, origin and favorable environmental conditions for *Coffea arabica* and *C. canephora*

Coffee, one of the most traded commodities in the world, is supported by *C. arabica* L. and *C. canephora* Pierre ex A. Froehner species [14]. It is estimated that the coffee chain of value generates a global income of *ca.* US$ 173,000 million [50], having as well great social implications. In fact, this tropical crop is grown in approximately 80 countries [51], and about 25 million farmers, mainly smallholders, depend on this highly labor-intensive crop [52], with a worldwide involvement of *ca.* 125 million people in the entire chain [53]. Brazil, Vietnam, Colombia, Indonesia, Ethiopia, India, Honduras and Uganda are the major coffee producers, for a world annual production of green coffee beans which has been increasing steadily in the last decades, being consistently near or above 9 million tons since 2011/2012 [54]. This supports over 2.5 billion cups of coffee consumed every day around the world [55], with promising prospects for increased consumption in the coming years, especially among young people in Asia.

The coffee plant is characterized as a perennial woody shrub that belongs to the Rubiaceae family. Although there are at least 125 species within the *Coffea* genus [56], *Coffea arabica* L. (Arabica coffee) and *Coffea canephora* Pierre ex A. Froehner (Robusta coffee) are responsible for approximately 99% of the world coffee production [23, 57], with the former accounting for ~65% of total coffee production [55, 58]. Besides differences in origin, these species present important ecological differences in plant traits, as well in bean chemical composition, among them aroma precursors. In fact, the levels of these compounds have implications on sensory attributes, namely on astringency, taste, aroma, and flavor after roasting. Such chemical composition is not only genetic related but also strongly depend on environmental conditions (e.g., soil, shade, temperature), bean maturation stage, and to agricultural management and post-harvest procedures [59–64].

C. arabica are originated from the tropical forests of Ethiopia, Sudan and Kenya, at altitudes of 1500–2800 m, annual averages air temperatures between 18 and 22°C, precipitation from 1600 to more than 2000 mm l distributed throughout the year, with a well-dry season (3–4 months), coinciding with the cold annual period. Currently, *C. arabica* coffee is grown in areas with cooler temperatures (18–23°C), at altitude mostly between 400 and 1200 m [7, 30, 65, 66], although cultivation up to 2000 m can be found in some countries in Central America. In contrast, *C. canephora* originated from the lowland forests of the Congo River basin, which extend to Lake Victoria in Uganda at altitudes up to 1200 m, are subjected to annual averages air temperatures between 23 and 26°C with minor fluctuations, and average precipitation exceeding 2000 mm distributed along 9–10 months [67–69]. Currently, cultivation occurs predominantly in lower altitude areas and higher temperatures, showing satisfactory development when the daily average temperature is above 22°C so that minimum is above 17°C and the average maximum air temperatures are below 31.5°C, with regular pattern of precipitation [70–74].

3. The impact of climate changes on coffee crop: warming and water scarcity

Coffee plants require both adequate water supply and optimal temperature, which are considered the most important environmental variables, since water and temperature-limited conditions cause negative impacts on growth, yield and productivity [14, 16, 30, 75]. Although in many coffee producing areas water scarcity occurs in the cooler season, climate modifications has increased the situations where low water availability and elevated temperature occur concomitantly under field conditions, which, as observed in other plants, will have the potential to exacerbate the limitations to the photosynthetic functioning [76].

In plants, photosynthesis and respiration are among the most sensitive metabolic processes to increasing temperatures [77]. High temperatures can cause protein denaturation and aggregation, increased production of reactive oxygen species [14], and ethylene synthesis [78]. Moreover, supra-optimal temperatures can reduce stomatal conductance and light energy use as well as alter thylakoid ultrastructure and diffusion of gas through mesophyll [15, 79–81] with a direct impact on net C gain. The latter will be even more amplified due to the increase of O_2 solubility in relation to CO_2 under higher temperatures, favoring the oxygenase activity of

RuBisCO over its carboxylation activity, thus increasing photorespiration rates [82, 83]. Altogether, this ultimately may lead to the decline in the availability of carbohydrates for energy supply as well as carbon skeletons to support plant growth [77]. Thus, warmer temperatures can affect crop yield at any time from sowing to grain maturity, but it is the time around flowering, when the number of grains per land area is established, and during the grain-filling stage, when the average grain weight is determined, that high temperatures causes major impacts on the final harvestable crop [9, 73, 84]. In addition, it causes a reduction in the production of leaves and consequently alters the photosynthetic activity [85].

Coffee trees presented a remarkable tolerance to temperatures relatively high (up to 37/30°C; day/night) when air humidity was maintained at 75%, occurring relevant physiological/biochemical impairments only at 42/34°C, associated, namely, to large activity reductions of RuBisCO and Ru5PK [35], despite large accumulation of RuBisCO transcripts [86]. The reported heat tolerance was related with increases in protective molecules, namely, enzyme and non-enzyme antioxidant molecules, heat shock protein 70 (HSP70) reinforcement, and altered gene expression [11, 86]. However, under field conditions, rising temperature may lead to increase in air vapor pressure deficit (VPD_{air}), what may result in decreased stomatal and canopy conductance in *Coffea* spp., due to a high sensitivity of coffee stomata to VPD_{air} values above 2 kPa [87–89]. In addition, elevated temperatures can contribute to a gradual increase in soil water depletion, particularity in areas lacking sufficient precipitation, resulting in water stress, which further exacerbates the adverse effects of high temperatures.

Stomatal closure is one of the first responses to water deficit in coffee plants, aiming at limiting water loss through transpiration flow. However, this directly decreases the CO_2 availability in the chloroplasts, reducing the photosynthetic rates [14]. In this context, irradiance reaching the chloroplasts may exceed the light energy needed to saturate photosynthesis, which in turn can lead to the formation of reactive oxygen species (ROS). ROS can cause oxidative damage to multiple cell and chloroplast components, namely to the D1 protein, lipids, RNA and DNA molecules, associated with increased cellular and metabolic disorders, resulting in cell death [47, 90, 91]. Moreover, ethylene synthesis often increases under drought stress conditions, promoting leaf senescence and slowing growth [10]. However, coffee plants display a noticeable metabolic plasticity to cope with environmental stresses [14, 51], as referred above for supra-optimal temperatures. Additionally, air [CO_2] enrichment improved both coffee antioxidant defense system and photosynthetic performance regardless of temperatures, but maintaining a relevant photosynthetic functioning at temperature as high as 42°C. This prevented an energy overcharge in the photosynthetic apparatus, eventually reducing the need for energy dissipation and PSII photoinhibition [11, 35].

Considering water stress, a large number of early studies have reported that coffee plants can cope with drought stress through morphological, biochemical, and physiological modifications [14], as discussed later in this chapter. However, prolonged drought events associated with elevated temperatures can lead to very severe conditions, with a general impact on cell metabolism, associated as well to increased oxidative stress, altogether resulting in intense defoliation and yield losses (**Figure 1**), although genotypic difference in stomatal sensitivity to water stress among *C. canephora* genotypes have been reported [43, 45]. Furthermore, drought

Figure 1. Intensive defoliation and yield losses due to prolonged severe drought (A), together with high temperatures (B) in *C. canephora* cultivations in Espírito Santo State, Brazil.

should be envisaged as contributing to a multidimensional stress, exacerbating the negative impacts of elevated irradiance and supra-optimal temperatures [13, 14, 42]. Therefore, drought-resistant coffee genotypes are a useful strategy for improving coffee performance in regions that are predicted to face moderate to severe drought [49].

Overall, drought-sensitive *C. canephora* genotypes show a shallow root system and ineffective stomatal control, whereas drought-resistant coffee genotypes show considerably deeper root system, the strengthening of antioxidant defense system, and higher stomata sensitivity to reduced water availability (both in soil and atmosphere) [43, 92]. Increased wood density reinforcing vessels and, in turn improving resistance to cavitation, was correlated with tolerance to hydraulic dysfunctions [45]. On the other hand, *C. canephora* genotypes with specific traits conferring drought tolerance generally show reduced yield under optimal environments conditions due to their increased stomata sensitivity to VPD_{air}. This is related to hydraulic limitations to water flow from roots to leaves [43, 45]. Therefore, coffee genotypes displaying increased phenotypic plasticity as, e.g., deep root system, substantial hydraulic conductance, intermediate stomatal control and strengthening of antioxidant defense system, could be used in regions which are predicted to face moderate water deficit, while drought-resistant genotypes could be used in regions predicted to face severe drought.

In addition to the traits outlined above, leaf size as well as canopy architecture should also be considered as important traits associated with drought tolerance. For example, although the leaf hydraulic conductivity (K_{leaf}) values found in *C. arabica* plants are typically low, probably linked to their native shade habitat [44, 93], *C. arabica* coffee genotypes with smaller leaves displayed higher vein density, higher K_{leaf}, increased gas exchange and reduced drought vulnerability [40, 44]. Drought tolerance was also found to be higher for *C. canephora* genotypes displaying smaller leaves [42]. In fact, it is known in other plants that smaller leaves allow for more rapid convective heat loss, resulting in lower transpiration and water loss likely due to smaller boundary layer [94]. Furthermore, a more compact crown structure may result in reduced VPD_{air} within the coffee canopy, decreasing the transpiration demand [14], besides allowing to increase plant density coupled with improved soil covering and reducing the negative impacts of elevated temperatures, and high wind speed on coffee trees. On the other hand, *C. arabica* genotypes displaying open architecture crown show high transpiration rates (as measured by the sap flow technique) depleting accessible soil water more rapidly [40]. Therefore, although the water use efficiency in coffee genotypes is associated with the hydraulic capacity of the soil and stem to supply the leaves with water [95], coffee traits linked to water safety, e.g., a more compact crown structure and to greater extent an effective stomatal control, seem to play an important role in drought tolerance.

A recent study by [48] reported that both drought-sensitive and drought-tolerant *C. canephora* genotypes showed a drought stress "memory," with plants exposed to multiple drought events showing better recovery than those submitted to drought events for the first time. This performance was mainly associated with substantial metabolic reprogramming, involving key processes such as photosynthesis, respiration, photorespiration, and the antioxidant system. In this sense, it would appear reasonable to suggest that multiple moderate water stress in coffee seedlings at nursery stage may improve to some extent the initial coffee performance under field conditions in areas prone to water scarcity.

4. Can elevated [CO_2] help the mitigation of the negative impacts of high temperature and water deficit?

Although climate models point CO_2 as the major greenhouse gas responsible for global warming due to its high accumulation rate in the atmosphere [6], the impacts of increased air [CO_2] at plant physiological and biochemical levels should not be neglected, namely in coffee metabolism [11, 31, 32, 35], as well in yield [36, 37].

The current [CO_2] in the atmosphere is still below the optimum for photosynthesis of C3 plants; therefore, leaf photosynthetic rates are predicted to increase in response to future increase in air [CO_2], due to increased carboxylase activity of RuBisCO [82, 83, 96]. This C-fertilization may eventually reinforce plant vigor (and the defense systems), which, in turn, could reinforce the plant ability to endure environmental stresses [97]. On the other hand, elevated CO_2 levels will especially benefit plants with strong sink capacity to use such increased amounts of photoassimilates. Otherwise, an accumulation of soluble sugars may occur which in turn will decrease the net photosynthetic rate through negative feedback mechanisms, that is, will provoke *downregulation* of photosynthesis, not allowing the plant to fully explore the positive effect of [CO_2] increase [83].

In the case of coffee, significant increases of net photosynthesis, between 34 and 49%, were observed for *C. canephora* (Clone 153) and *C. arabica* (Icatu and IPR 108) genotypes [31], when comparing plants grown subjected to elevated [CO_2] (700 $\mu L\ L^{-1}$) or normal [CO_2] (380 $\mu L\ L^{-1}$) under environmental controlled conditions. Furthermore, under such high [CO_2], plants also showed a better water-use efficiency, reinforcement of photosynthetic components and increased activity of key enzymes involved in photosynthesis and respiration, without noticeable leaf sugar accumulation. Therefore, these coffee genotypes were able to cope with enhanced [CO_2], maintaining the consumption of photosynthates and regeneration of RuBP associated with continuous investment in vegetative and reproductive structures. The evidence of improved coffee performance under enhanced [CO_2] was further obtained with other *C. arabica* genotypes (Obatã IAC 1669–20 and Catuaí Vermelho IAC 144) under field conditions using free-air CO_2 enrichment (FACE) system, showing increased photosynthesis and decreased photorespiration, without changes in stomatal and mesophyll conductance, for an air [CO_2] of 550 $\mu L\ L^{-1}$ [33]. Additionally, coffee plants grown under elevated [CO_2] were more vigorous, with increased leaf area, growth rate at height and stem diameter, showing as well increased grain yield by 14.6 and 12.0% for Catuaí Vermelho 144 and Obatã IAC 1669–20, respectively, [8, 32], although average yield increases of 28% were also reported after three harvests [37] when compared to plant grown at ambient [CO_2]. Another study also demonstrated that coffee trees grown under 550 $\mu L\ CO_2\ L^{-1}$ presented increase in photosynthesis of leaves from upper and lower canopy layers, inhibition of photorespiration, and no apparent sign of photosynthetic *downregulation*, when compared to plants grown under ambient [CO_2] (390 $\mu L\ L^{-1}$) [98]. Finally, recent studies based on modeling approaches accounting with high air [CO_2] positive impact reported that coffee yield losses associated mostly with high temperatures can be offset by the CO_2 fertilization effect, with a probably yield increase by 2040–2070 [36], or 2050, particularly at higher altitudes [37].

The simultaneous occurrence of various environmental constraints is the most common situation under field conditions, and therefore, it has been argued that a positive plastic response from plant experiencing a single stress can be increased, canceled or even reverted under the combined action of multiple stresses [6]. Regarding the coffee plant, responses to the combined effects of increased [CO_2] and supra-optimal air temperature started to be investigated quite recently, whereas the simultaneous exposure to elevated [CO_2], heat and water deficit have never been studied. The exposure to increased air [CO_2] revealed interesting implications to plant physiological response to supra-optimal conditions. This was the case in both *C. arabica* (cvs. Icatu and IPR 108) and *C. canephora* cv. Conilon Clone 153 plants exposed to elevated [CO_2] and temperatures up to 42°C [11, 34, 35]. Notably, a remarkable heat tolerance was observed up to 37/30°C (day/night) irrespective of air [CO_2]. The tolerance (and high physiological performance) to such temperature was somewhat surprising as it is above what is traditionally accepted to be tolerated by coffee plant [35]. Furthermore, enhanced [CO_2] greatly mitigated the negative impact of the temperature, especially at 42/34°C, with higher water-use efficiency (WUE) at moderately higher temperature (31/25°C). Increased CO_2 was observed to strengthen the photosynthetic apparatus, improving light energy use and biochemical functioning. These results were linked to the maintenance or increase in the content of several protective molecules (neoxanthin, lutein, β-carotene, α-tocopherol, heat shock protein-HSP70, raffinose), the activity of antioxidant enzymes (superoxide dismutase, SOD; ascorbate peroxidase, APX, glutathione reductase, GR; catalase, CAT) and the upregulation of some genes related to stress-protective molecules (ELIP, HSP70, Chaperonin 20 and 60), and antioxidant enzymes (*CAT, CuSOD2, APX Cyt, APX Chl*) [11]. In the same experiments, overall leaf mineral macro- and microelement contents have remained within a range that could be considered largely adequate for coffee plants, with no changes in macronutrient profile (N > K > Ca > Mg > S > P), that is, satisfactory mineral content was maintained in the context of warming, under high [CO_2] [34].

Climate changes are also predicted to affect intra- and inter-annual rainfall patterns, and the decrease in precipitation amounts in conjunction with increased air temperature may reduce net photosynthesis at current [CO_2]. Still, under increased air [CO_2], a partial relief of negative impacts of water deficit may occur [99]. Indeed, arabica coffee plants grown under severe drought conditions and increased biotic pressure showed strategies which allow the maintenance of structural and physiological integrity in the fourth period of winter growth [98]. This occurs because of the dichotomous responses of net photosynthesis and stomatal conductance to high [CO_2], which lead to improved WUE, reducing soil moisture depletion during periods of drought [9]. Studies by [10, 76] on *Agropyron cristatum* L. and *Perilla frutescens* var. japonica Hara, respectively, reported positive results of elevated-CO_2 mitigation of drought stress, verifying increase in photosynthetic capacity and decrease in stomatal conductance with lower transpiration rates. Consequently, increased intrinsic water-use efficiency (WUEi) and total water-use efficiency (WUEt) were observed. Furthermore, high [CO_2] can also alleviate oxidative stress conditions, and photoinhibition status, likely associated to a higher photosynthetic functioning (as also observed for high temperatures [11]), even under significant stomatal closure. Altogether such responses may result in improved tolerance to drought stress, as found in other plants [6, 10, 12]. Nevertheless, it is important to note that under severe drought, such positive results might not be obtained, and that mitigation associated with high [CO_2] does not always occur [6].

In addition to the positive effects on the impacts of abiotic stresses, elevated [CO_2] can also reduce to some extent the incidence and severity of coffee pests and diseases. In fact, decrease in leaf rust (*Hemileia vastatrix*) severity, number of lesions, leaf area injured, number of sporulating lesions, percentage of damaged leaf area and area under disease progress were observed in *C. arabica* cv. Catuaí IAC 144 grown under elevated [CO_2] [8]. Reduced incidence of leaf miner (*Leucoptera coffeella*) during periods of high infestation was also observed at elevated [CO_2] [32].

In summary, enhanced [CO_2] can have a positive mitigation effect on the negative impacts of high temperature and, probably, low water availability, as well as by reducing the severity of some pests and diseases. However, since responses are highly species (and even cultivar) dependent, it is urgent to implement long-term studies in coffee considering single and, especially, combined stresses, with the simultaneous exposure to elevated [CO_2], supra-optimal temperatures and drought, relating them to phenological stages (e.g., flowering), therefore, to increase knowledge on this crop in a context of climate changes.

5. Mitigating the impacts of climate changes through management practices

To promote crop sustainability in the context of climate changes and global warming, adaptation and mitigation measures must be implemented. Regarding adaptation, plant screening and breeding are essential to provide new improved and stress-tolerant genotypes, but their implementation are somewhat delayed due to the time needed to obtain new varieties. As an example, the use of improved genotypes with an optimized architecture is a valuable tool. It is known that small-size plants, with denser canopies, are prone to display lower transpiration rates [13, 14]. Additionally, plants with larger and deeper root systems would have an ability to explore increased soil volumes, reaching water resources that other plants with a more superficial root system do not [14]. Still, several years will be needed until such new genotypes can be available and, therefore, ready-to-use strategies should be implemented, namely those regarding an effective mitigation of the environmental negative impacts on the actual cropped genotypes. This can be even more important when dealing with tree crops that have a productive lifespan of several years or decades, as it is the case of coffee, which can last for more than 30 years [18].

A significant range of management techniques can be used to minimize the impact of different stresses that can affect the performance of agricultural systems. For coffee crop, several different agronomic tools stand to that purpose, e.g., the use of shade systems with tree species, as well as other intercropping associations, to improve an efficient water use and minimize warming at the plant level, maintaining a more suitable microenvironment concerning both temperature and air humidity. Improved soil covering with other intercropped species, and terracing under conditions of significant slopes, are also quite useful techniques to minimize soil water loss (or to increase its infiltration), therefore, helping to maintain water resources available to the plants for longer periods.

5.1. Fertilization management under high air [CO_2] and warming conditions

Minerals have a wide number of roles in plant cell. Therefore, as in other plants, an adequate mineral fertilization is recognized as crucial to allow the triggering of acclimation mechanisms in face of environmental constraints in the coffee plant. This is the case of nitrogen (N) supply, which is of utmost importance to allow the recovery from high irradiance impact, through the triggering of repair mechanism, and the reinforcement of leaf defense mechanisms, including the control of highly reactive molecules of chlorophyll and oxygen, whose production is exacerbated under high irradiance/full sun exposure [100–102]. Additionally, the presence of adequate contents of other minerals allows the plant to maintain high metabolic performance due to their specific roles. For instance, copper, iron and manganese, which were shown to promote the activities of, respectively, superoxide dismutase, ascorbate peroxidase, and photosystem II under cold exposure [103], as well as calcium, which is essential to the stabilization of chlorophyll and the maintenance of photochemical efficiency at PS II level [104].

Changes in mineral contents may affect plant development, but may also have other important consequences, namely as regards the quality of agricultural products for food and feed, herbivory, litter decomposition rates, etc. [105, 106]. It is known that mineral contents often decline in the leaf biomass under high air [CO_2] conditions. This was related to higher growth rates, accumulation of non-structural sugars (mainly starch), lower transpiration rates, or to changes in the nutrient allocation patterns under enhanced air [CO_2] [107–109] This mineral "dilution" effect on leaves can affect the photosynthetic apparatus (e.g., through N, S and Fe), enzyme activity (e.g., through K, P, Mn and Fe), alters redox reactions (e.g., through Fe, Zn and Cu), and modifies the structural integrity of chloroplast membranes (e.g., B) [105, 110–113]. However, this so called "dilution effect" may frequently reflect qualitative physiological changes rather than a lack of nutrients [108], since in many cases, these plants did not present mineral nutrition disturbances. This seems just to be the case in *Coffea* spp., since it was observed that under adequate temperature, long-term exposure to enhanced [CO_2] (700 μL L^{-1}) net photosynthetic rate was increased by between 40 and 49% [31], concomitantly to a moderate mineral reduction that ranged from 7 to 25% in N, Mg, Ca, Fe in *C. canephora* cv Conilon Clone 153, and in N, K and Fe in *C. arabica* cv. Icatu [34].

Most important was also the observation that contents (on a per leaf mass basis) of several minerals increased under supra-optimal temperatures, largely offsetting the dilution effect observed under control temperature (25°C), keeping the large majority of minerals and their ratios within a range that is considered adequate, therefore, suggesting that coffee plant can maintain its mineral balance in a context of climate changes and global warming [34]. Even so, taking into account the importance of mineral dynamics to virtually all biological processes, studies under field conditions must be implemented to better understand the possible CO_2 implications for coffee fertilizer management in a context of climate changes and global warming in a near future.

5.2. Reducing irradiance at the leaf level

Both *C arabica* and *C. canephora* have been cultivated under full sunlight in many regions around the world, particularly in Brazil. In fact, coffee plant can successfully adjust its photosynthetic metabolism to high light conditions, namely if adequate mineral nutrition is provided [100–102]. Effective acclimation to other environmental constraints (e.g., cold, heat, drought) was also reported [14]. Such acclimation ability depends on the presence and/or reinforcement of several mechanisms, among them leaf antioxidants, and qualitative modifications on the lipid matrix of cell membranes, particularly in the chloroplast. This allows the plant to maintain high metabolic activity, namely as regards the photosynthetic pathway, depending on stress severity and on species and genotype capabilities [11, 41, 57, 93, 101, 114]. However, these coffee species have evolved and grow naturally under shaded understory [14, 68, 69]. Not surprisingly, *Coffee* sp. presents some leaf traits usually associated with shade plants, namely low light saturating point (*ca.* 500 μmol m^{-2} s^{-1}) [115], therefore, quite below the irradiance values occurring under field conditions. This increases the probability of photoinhibition under high solar radiation [13, 14, 100, 116, 117]. Taking into account predictions of a global warming and lower water availability along the present century, the implementation of coffee cultivation under shaded conditions (e.g., under agro-forestry systems) may be recommended as a cultural management practice to alleviate the combined impacts of drought and elevated temperatures [118], while improving nutrient cycling, soil fertility and soil organic matter accumulation [119–122]. Additionally, shade crops can improve ecological aspects including increasing bio-diversity of flora and fauna [123, 124].

Traditionally, coffee trees grown under shaded conditions show reduced yield, since shade trees may compete with coffee for essential requirements such as light, water and nutrient depending on tree density [13, 119, 125], with less nodes per branch and fewer flowers at existing nodes must be also considered. Additionally, coffee plants show limited light distribution within their own canopies [88], thus leading to the further reduction of the light availability at whole canopy scale. However, increased light-use efficiency can compensate the low availability of photosynthetically active solar radiation in coffee trees grown under shaded conditions [126]. Also, shade trees can increase the proportion of diffuse light under their canopy by 60–90%, what may lead to increased penetration of radiation inside the coffee canopy [126]. In fact, *C. canephora* Clone 02 (clonal variety "EMCAPA 8111" [127]), grown under an irradiance retention of 70% promoted by Australian cedar (*Toona ciliata* M. Roem) in southeastern region of Brazil showed similar yield to unshaded counterparts, although for a study considering only one crop season [128] (**Figure 2**). Similar yield and leaf nutrient content were also found in shaded *C. canephora* cv. Verdebras G35 plants intercropped with rubber trees (*Hevea brasiliensis* (Willd. ex A. Juss.) Müll. Arg.) in the same region, with a reduction of *ca*. 70% in total irradiation [129], while similar yield were reported for *C. arabica* cv. Caturra intercropped with *Erythina poeppigiana* (reduction of *ca*. 70% in total irradiation) in the central Valley of Costa Rica [126] and in six *C. arabica* genotypes shaded by *E. verna* and *Musa* sp. (shade up to 80%) [130].

As referred above, coffee trees show increased stomatal sensitivity to VPD_{air}, so that increase in air temperature and/or decrease in air relative humidity (RH) can result in reduced stomatal

Figure 2. *Coffea canephora* cv. Conilon under shading conditions promoted by A) Australian cedar (*Toona ciliate* M. Roem. var. Australis), B) papaya (*Carica papaya* L.), C) rubber tree (*Hevea brasiliensis* Willd. ex A. Juss), and D) African mahogany (*Khaya* spp.), in northern Espírito Santo state, Brazil.

aperture. In this sense, shaded systems with trees, including rubber [129] and Australian cedar [128], can reduce air temperature, maintain higher air humidity, and decrease low wind speed near the coffee plants, thus resulting in decreased VPD_{air} between the leaf and the atmosphere, and a lower water loss through transpiration [13]. Therefore, shade will promote a better WUE, reducing plant transpiration and soil evaporation, while contributing to improve plant physiological performance [117].

In addition to the impacts on photosynthetic machinery, rising temperature causes increases in plant respiration rates, mainly associated with "maintenance respiration" to support protein turnover and to maintain active ions transport across the membrane [81]. Recent studies have reported decreases of 2 up to 6°C in air temperature surrounding coffee canopy under shaded condition [125, 128, 129, 131]. Such reduction in air temperature can therefore reduce maintenance respiration [126], as *C. arabica* cv. Caturra plants in the Central Valley of Costa Rica that showed a 40% decrease in peak maintenance respiration under a 4°C decrease in maximum temperature [125].

Coffee growers need to obtain high yields, while maintaining bean quality in order to guarantee their income. Rising temperature may decrease coffee bean yields due to bud abortion or development of infertile flowers, particularly when associated with prolonged dry periods [65]. Additionally, increased temperature may accelerate fruit maturation and ripening, reducing the accumulation of sucrose and altering the content of several compounds that are known precursors of taste, flavor and aroma after roasting [15, 60, 62, 64]. Shade trees may provide a milder microclimate, attenuating temperature rise on coffee beans, and by lowering air temperature close to the coffee plant can extend the maturation period so that the bean filling period will be enlarged [132, 133], what can contribute to higher sucrose accumulation.

Besides the importance of shade in reducing thermal stress, other important benefits arise as well. For instance, coffee trees grown under full sunlight show a typical biennial pattern, e.g., during one crop season, a heavy fruit load will constitute a major sink at the expense of new leaves and branches, reducing productivity in the following year [134]. Moreover, high fruits load may result in reduced bean size due to the carbohydrate competition among berries during bean filling [133]. In this sense, depending on density, shade trees can reduce coffee flowering intensity, resulting in a better coffee bean quality, as well as in higher yield stability along the years. Although the central purpose of coffee cultivation under shaded conditions is alleviating the impacts of both high irradiances and supra-optimal temperatures, it is worth to mention that cultivation of trees of economic importance, such as *Inga* sp. [125], Australian cedar [128], rubber tree (**Figure 2**) [129], can constitute important complementary sources of income to coffee farmers.

The application of kaolin particles can also reduce the irradiance at leaf surface, increasing radiation reflections, and, consequently decreasing leaf temperature [135]. Kaolin particle film can as well improve light distribution inside the canopy, leading to increase in photosynthetic rates, increasing crop water use efficiency at whole-canopy scale, as reported for apple (*Malus sylvestris*) [136, 137] and grapevine (*Vitis vinifera* L.) [137]. Moreover, kaolin particle film protected apple fruits from damage caused by excessive heat linked to high light conditions, besides avoiding the direct impacts of ultraviolet radiation on fruits as well [135]. Additionally, some works have demonstrated that particle film technology can alleviate the negative impacts of water stress, particularly associated with increase in light reflection and decrease in canopy temperature [137, 138]. In coffee, kaolin particle film was observed to increase C-assimilation and bean yield, linked to improved light distribution within the canopy, since sunlight is essential to floral initiation [139], and can, therefore, constitute a promising alternative technique to reduce the thermal energy at leaf level.

Considering the effects of supra-optimal temperatures, high density planting system can alleviate the negative impacts of heat stress, because under such conditions, the air surrounding the coffee plants becomes more humid due to plant transpiration and low wind speed, decreasing VPD_{air} [14]. Additionally, in areas facing strong winds, the use of windbreaks or tree shelters is recommended as both can avoid an extensive removal of boundary layer, leading to decreased demand for water from the atmosphere. However, under high density planting systems, coffee crop management through pruning is fundamental for renewal, revitalizing and yield stability in coffee plantations [140], what can improve soil coverage.

5.3. Soil covering and terracing

The distance between coffee rows allows for growth of other plants, which may compete for water and nutrients, depending on species involved. Overall, weed control aims at removing the invasive plants, exposing soil to intense solar radiation which can result in increase in water evaporation directly from the soil as well as facilitating the surface water runoff, leading to erosion losses, especially in areas with a pronounced slope. Depending on weed species, invasive plants are allowed to grow naturally between coffee rows without any management strategy. Although such plants may reduce erosion losses and direct solar radiation, as well as improve the infiltration of water into the soil stratum [141], they lose water during the day through transpiration, decreasing soil moisture [142]. Therefore, weed management strategies (for example, cut using a mower) can contribute for organic matter accumulation and, in turn, increase the water retention capacity of the soil, improving water productivity.

Also, the use of some leguminous species, correctly managed between coffee rows, can protect the soil, providing N to the coffee plants. Furthermore, soil coverage with herbaceous plants between coffee rows increases soil moisture and reduces both soil temperature and weed incidence, improves the physical and chemical soil properties [143, 144], promotes water infiltration, reduces rainfall impact and erosion, stimulates microbial activity, and improves organic matter in the soil [145]. Improved ground cover can be further obtained from weeds control, and by keeping biomass from coffee plants pruning, a common practice used to promote crop productivity [140] and soil microbiota diversity.

Coffee straw/husks, a by-product generated during coffee processing and discarded in many farms, can also be used for soil covering, reducing water losses through soil evaporation. In addition, coffee straw/husks can provide essential macro and micronutrients, namely N, P, K, Ca, Mg, S, Fe, B, Mn, Zn and Cu [72], lowering the need of chemical fertilization regarding these nutrients, and increasing coffee yield up to 25% [146]. Moreover, these coffee by-products can improve the soil physical associated with increase in CTC and soil pH [147], and inhibit seed germination of many weed species such as *Amaranthus retroflexus, Bidens pilosa, Cenchrus echinatus* and *Amaranthus spinosus* [148]. Therefore, coffee straw/husks can increase soil water retention and reduce to some extent costs associated with weed managements and fertilizers.

Other strategies for areas with a high slope are terracing, contour plowing terrace and rectangular ditches. Such practices contribute for preventing rapid surface runoff, allowing rain water to percolate into the soil, contributing for soil conservation [149–151]. Therefore, the establishment of terraces, although expensive, could constitute a worthwhile alternative to reduce water losses through runoff and soil erosion, while promoting infiltration [152]. Rain water storage in reservoirs should also be implemented. This will allow future water use during periods of negligible rainfall, constituting an important mitigation strategy to avoid drought stress. Therefore, increasing the water retention/storage capability in the farm can delay or even prevent coffee water stress.

6. Future perspectives

Climate changes are expected to negatively affect the coffee crop, causing serious social and economic impacts. Supra-optimal temperatures and water scarcity may decrease coffee yields and some studies state that these stresses are already occurring in some coffee-growing countries. However, coffee plants show a potential ability to cope with several environmental stresses and enhanced [CO_2] can improve such ability and mitigate to some extent the negative impacts of supra-optimal temperatures. Even so, some mitigation strategies will be necessary to alleviate the impacts of elevated temperature and/or drought stress on coffee trees. We have reviewed some strategies that can be implemented depending on main environmental stresses occurring in specific regions, such as those based on coffee traits (root systems, size leaf, canopy architecture and stomatal sensitivity) and crop management (nutrient managements and pruning system), as well as those aiming at reducing excessive light at coffee tree level (shaded systems, kaolin-based particle film and plant density), and at improving soil water retention (soil covering and terracing). Notably, however, a single mitigation strategy may not be enough to face severe stress conditions; thus, multiple strategies should be undertaken.

Future studies considering simultaneous exposure to the main environmental stresses (e.g., high temperatures and drought), taking into account as well elevated [CO_2], will be necessary to elucidate the mechanisms underlying plasticity and vulnerability of coffee plants under conditions that are expected to occur in the fields in a near future. Such studies are a fundamental basis for plant breeders to obtain new/more adapted genotypes. Finally, these strategies appear to be useful tools toward maintaining the coffee chain production.

Acknowledgements

This work was partly supported by contribution of national funds from Fundação para a Ciência e a Tecnologia through the research units UID/AGR/04129/2013 (LEAF) and UID/GEO/04035/2013 (GeoBioTec). Brazilian funding from CAPES (grant PDSE: 88881.132375/

2016-01, Danielly Dubberstein), CNPq and Fapemig (fellowships to F.M. DaMatta, F. Partelli, E. Campostrini and W.P. Rodrigues-PDJ438817/2016-8) are also greatly acknowledged.

Author details

Danielly Dubberstein[1,2†], Weverton P. Rodrigues[3†], José N. Semedo[4,6], Ana P. Rodrigues[5], Isabel P. Pais[4,6], António E. Leitão[2,6], Fábio L. Partelli[7], Eliemar Campostrini[3], Fernando Reboredo[6], Paula Scotti-Campos[4,6], Fernando C. Lidon[6], Ana I. Ribeiro-Barros[2,6], Fábio M. DaMatta[8] and José C. Ramalho[2,6*]

*Address all correspondence to: cochichor@mail.telepac.pt

1 Centro de Ciências Agrárias e Engenharias (CCAE), Universidade Federal do Espírito Santo (UFES), Alegre, ES, Brazil

2 Linking Landscape, Environment, Agriculture and Food (LEAF), Instituto Superior de Agronomia (ISA), Universidade de Lisboa (ULisboa), Oeiras, Portugal

3 Setor Fisiologia Vegetal, Centro de Ciências e Tecnologias Agropecuárias, Universidade Estadual Norte Fluminense – Darcy Ribeiro (UENF), RJ, Brazil

4 Unidade de Investigação em Biotecnologia e Recursos Genéticos (UIBRG), Instituto Nacional de Investigação Agrária e Veterinária, I.P. (INIAV), Oeiras, Portugal

5 Forest Studies Center (CEF), Instituto Superior de Agronomia (ISA), Universidade de Lisboa (ULisboa), Lisboa, Portugal

6 Unidade de Geobiociências, Geoengenharias e Geotecnologias (GeoBioTec), Faculdade de Ciências e Tecnologia (FCT), Universidade NOVA de Lisboa (UNL), Monte de Caparica, Portugal

7 Centro Universitário Norte do Espírito Santo (CEUNES), Dept. Ciências Agrárias e Biológicas (DCAB), Universidade Federal Espírito Santo (UFES), São Mateus, ES, Brazil

8 Department of Biologia Vegetal, Universidade Federal Viçosa (UFV), Viçosa, MG, Brazil

† These authors contributed equally.

References

[1] IPCC. Climate change 2014: Mitigation of climate change. In: Edenhofer O, Pichs-Madruga R, Sokona Y, Farahani E, Kadner S, Seyboth K, Adler A, Baum I, Brunner S, Eickemeier P, Kriemann B, Savolainen J, Schlömer S, von Stechow C, Zwickel T, Minx JC, editors. Contribution of Working Group III to the Fifth Assessment Report of the Intergovernmental Panel on Climate Change. Cambridge, United Kingdom and New York, NY, USA: Cambridge University Press; 2014

[2] Van Beek CL, Meerburg BG, Schils RLM, Verhagen J, Kuikman PJ. Feeding the world's increasing population while limiting climate change impacts: Linking N_2O and CH_4 emissions from agriculture to population growth. Environmental Science & Policy. 2010;**13**:89-96. DOI: https://doi.org/10.1016/j.envsci.2009.11.001

[3] Rahn E, Läderach P, Baca M, Cressy C, Schroth G, Malin D, van Rikxoort H, Shriver J. Climate change adaptation, mitigation and livelihood benefits in coffee production: Where are the synergies? Mitigation and Adaptation Strategies for Global Change. 2013;**19**:1119-1137. DOI: 10.1007/s11027-013-9467-x

[4] Beach RH, Cai Y, Thomson A, Zhang X, Jones R, Mccarl BA, Crimmins A, Martinich J, Cole J, Ohrel S, Deangelo B, Mcfarland J, Strzepek K, Boehlert B. Climate change impacts on us agriculture and forestry: Benefits of global climate stabilization. Environmental Research Letters. 2015;**10**:095004. DOI: 10.1088/1748-9326/10/9/095004

[5] Tack J, Barkley A, Nalley LL. Effect of warming temperatures on US wheat yields. Proceedings of the National Academy of Sciences of the United States of America. 2015;**112**:6931-6936. DOI: 10.1073/pnas.1415181112

[6] Xu Z, Shimizu H, Yagasaki Y, Ito S, Zheng Y, Zhou G. Interactive effects of elevated CO_2, drought, and warming on plants. Journal of Plant Growth Regulation. 2013;**32**:692-707. DOI: 10.1007/s00344-013-9337-5

[7] Zullo Jr J, Pinto HS, Assad ED, Ávila AMH. Potential for growing Arabica coffee in the extreme south of Brazil in a warmer world. Climatic Change. 2011;**109**:535-548. DOI: 10.1007/s10584-011-0058-0

[8] Tozzi FRO, Ghini R. Impact of increasing atmospheric concentration of carbon dioxide on coffee leaf rust and coffee plant growth. Pesquisa Agropecuária Brasileira. 2016;**51**:933-941. DOI: 10.1590/S0100-204X2016000800005

[9] DaMatta FM, Grandis A, Arenque BC, Buckerigde MS. Impacts of climate changes on crop physiology and food quality. Food Research International. 2010;**43**:1414-1423. DOI: https://doi.org/10.1016/j.foodres.2009.11.001

[10] Lee SH, Woo SY, Je SM. Effects of elevated CO_2 and water stress on physiological responses of *Perilla frutescens* Var. *japonica* HARA. Plant Growth Regulation. 2015;**75**:427-434. DOI: 10.1007/s10725-014-0003-0

[11] Martins MQ, Rodrigues WP, Fortunato AS, Leitão AE, Rodrigues AP, Pais IP, Martins LD, Silva MJ, Reboredo FH, Partelli FL, Campostrini E, Tomaz MA, Scotti-Campos P, Ribeiro-Barros AI, Lidon FJC, DaMatta FM, Ramalho JC. Protective response mechanisms to heat stress in interaction with high [CO_2] conditions in *Coffea* spp. Frontiers in Plant Science. 2016;**29**:947-964. DOI: https://doi.org/10.3389/fpls.2016.00947

[12] AbdElgawad H, Zinta G, Beemste GTS, Janssens IA, Asard H. Future climate CO_2 levels mitigate stress impact on plants: Increased defense or decreased dhallenge? Frontiers in Plant Science. 2016;**7**:556-562. DOI: https://doi.org/10.3389/fpls.2016.00556

[13] DaMatta FM. Ecophysiological constraints on the production of shaded and unshaded coffee: A review. Field Crops Research. 2004;**86**:99-114. DOI: https://doi.org/10.1016/j.fcr.2003.09.001

[14] DaMatta FM, Ramalho JC. Impacts of drought and temperature stress on coffee physiology and production: A review. Brazilian Journal of Plant Physiology. 2006;**18**:55-81. DOI: http://dx.doi.org/10.1590/S1677-04202006000100006

[15] Santos CAFD, Leitão AE, Pais IP, Lidon FC, Ramalho JC. Perspectives on the potential impacts of climate changes on coffee plant and bean quality. Emirates Journal of Food and Agriculture. 2015;**27**:152-163. DOI: 10.9755/ejfa.v27i2.19468

[16] Ovalle-Rivera O, Läderach P, Bunn C, Obersteiner M, Schroth G. Projected shifts in *Coffea arabica* suitability among major global producing regions due to climate change. PLoS One. 2015;**10**:e0124155. DOI: 10.1371/journal.pone.0124155

[17] Zullo Jr J, Pinto HS, Assad ED. Impact assessment study of climate change on agricultural zoning. Meteorological Applications. 2006;**13**:69-80. DOI: 10.1017/S135048270600257X

[18] Bunn C, Läderach P, Rivera OO, Kirschke D. A bitter cup: Climate change profile of global production of Arabica and Robusta coffee. Climatic Change. 2015;**129**:89-101. DOI: 10.1007/s10584-014-1306-x

[19] Magrach A, Ghazoul J. Climate and pest-driven geographic shifts in global coffee production: Implications for forest cover, biodiversity and carbon storage. PLoS One. 2015;**10**:e0133071. DOI: https://doi.org/10.1371/journal.pone.0133071

[20] Gay C, Estrada F, Conde C, Eakin H, Villers L. Potential impacts of climate change on agriculture: A case of study of coffee production in Veracruz, Mexico. Climatic Change. 2006;**79**:259-288. DOI: 10.1007/s10584-006-9066-x

[21] Schroth G, Laderach P, Dempewolf J, Philpott S, Haggar J, Eakin H, Castillejos T, Moreno JG, Pinto LS, Hernandez R, Eitzinger A, Ramirez-Villegas J. Towards a climate change adaptation strategy for coffee communities and ecosystems in the sierra Madre de Chiapas, Mexico. Mitigation and Adaptation Strategies for Global Change. 2009;**14**:605-625. DOI: 10.1007/s11027-009-9186-5

[22] Craparo ACW, Van Asten PJA, Läderach P, Jassogne LTP, Grab SW. *Coffea arabica* yields decline in Tanzania due to climate change: Global implications. Agricultural and Forest Meteorology. 2015;**207**:1-10. DOI: https://doi.org/10.1016/j.agrformet.2015.03.005

[23] Davis AP, Gole TW, Baena S, Moat J. The impact of climate change on indigenous arabica coffee (*Coffea arabica*): Predicting future trends and identifying priorities. PLoS One. 2012;**7**:e47981. DOI: https://doi.org/10.1371/journal.pone.0047981

[24] Jha S, Bacon CM, Philpott SM, Méndez VE, Läderach P, Rice RA. Shade coffee: Update on a disappearing refuge for biodiversity. Bioscience. 2014;**64**:416-428. DOI: https://doi.org/10.1093/biosci/biu038

[25] Pezzopane JRM, Castro FS, Pezzopane JEM, Bonomo R, Saraiva GS. Climatic risk zoning for Conilon coffee in Espirito Santo, Brazil. Revista Ciência Agronômica. 2010;**41**:341-348. DOI: 10.1590/S1806-66902010000300004

[26] van der Vossen H, Bertrand B, Charrier A. Next generation variety development for sustainable production of arabica coffee (*Coffea arabica* L.): A review. Euphytica. 2015;**204**:243-256. DOI: 10.1007/s10681-015-1398-z

[27] Villers L, Arizpe N, Orellana R, Conde C, Hernandez J. Impacts of climatic change on coffee flowering and fruit development in Veracruz, México. Interciencia. 2009;**34**:322-329. DOI: 0378-1844/09/05/322-08 $ 3.00/0

[28] Conab. Acompanhamento da safra brasileira de café, v. 3 – Safra 2016, n. 4 – Quarto Levantamento, Brasília. 2016. pp. 1-77

[29] Moat J, Williams J, Baena S, Wilkinson T, Gole TW, Challa ZK, Demissew S, Davis AP. Resilience potential of the Ethiopian coffee sector under climate change. Nature Plants. 2017;**3**:art17081. DOI: 10.1038/nplants.2017.81

[30] Camargo MBP. The impact of climatic variability and climate change on arabica coffee crop in Brazil. Bragantia. 2010;**69**:239-247. DOI: http://dx.doi.org/10.1590/S0006-87052010000100030

[31] Ramalho JC, Rodrigues AP, Semedo JN, Pais IP, Martins LD, Simões-Costa MC, Leitão AE, Fortunato AS, Batista-Santos P, Palos IM, Tomaz MA, Scotti-Campos P, Lidon FC, DaMatta FM. Sustained photosynthetic performance of *Coffea* spp. under long-term enhanced [CO_2]. PLoS One. 2013;**8**:e82712. DOI: 10.1371/journal.pone.0082712

[32] Ghini R, Torre-Neto A, Dentzien AFM, Gerreiro-Filho O, Iost R, Patrício FRA, Prado JSM, Thomaziello RA, Bettiol W, Damatta FM. Coffee growth, pest and yield responses to free-air CO_2 enrichment. Climatic Change. 2015;**132**:307-320. DOI: 10.1007/s10584-015-1422-2

[33] DaMatta FM, Godoy AG, Menezes-Silva PE, Martins SCV, Sanglard LMVP, Morais LE, Torre-Neto A, Ghini R. Sustained enhancement of photosynthesis in coffee trees grown under free-air CO_2 enrichment conditions: Disentangling the contributions of stomatal, mesophyll, and biochemical limitations. Journal of Experimental Botany. 2016;**67**:341-352. DOI: 10.1093/jxb/erv463

[34] Martins LD, Tomaz MA, Lidon FC, DaMatta FM, Ramalho JC. Combined effects of elevated [CO_2] and high temperature on leaf mineral balance in *Coffea* spp. plants. Climatic Change. 2014;**126**:365-379. DOI: 10.1007/s10584-014-1236-7

[35] Rodrigues WP, Martins MQ, Fortunato AS, Rodrigues AP, Semedo JN, Simões-Costa MC, Pais IP, Leitão AE, Colwell F, Goulao L, Máguas C, Maia R, Partelli FL, Campostrini E, Scotti-Campos P, Ribeiro-Barros AI, Lidon FC, Damatta FM, Ramalho JC. Long-term elevated air [CO_2] strengthens photosynthetic functioning and mitigates the impact of supra-optimal temperatures in tropical *Coffea arabica* and *C. canephora* species. Global Change Biology. 2016;**22**:415-431. DOI: 10.1111/gcb.13088

[36] Verhage FYF, Antem NPR, Sentelhas PC. Carbon dioxide fertilization offsets negative impacts of climate change on Arabica coffee yield in Brazil. Climatic Change. 2017;**144**:671-685. DOI: 10.1007/s10584-017-2068-z

[37] DaMatta FM., Rahn E, Laderach P, Ghini R, Ramalho JC. Why could the coffee crop endure climate change and global warming to a greater extent than previously estimated? Clim Change, 2018. (submitted)

[38] Nunes MA. Water relations in coffee: Significance of plant water deficits to growth and yield: A review. Journal of Coffee Research. 1976;**6**:4-21. Available at: http://www.sidalc.net/cgi-bin/wxis.exe/?IsisScript=orton.xis&method=post&formato=2&cantidad=1&expresion=mfn=032673

[39] Gutiérrez MV, Meinzer FC, Grantz DA. Regulation of transpiration in coffee hedgerows: Covariation of environmental variables and apparent responses of stomata to wind and humidity. Plant Cell Environment. 1994;**17**:1305-1313. DOI: 10.1111/j.1365-3040.1994.tb00532.x

[40] Tausend PC, Goldstein G, Meinzer FC. Water utilization, plant hydraulic properties and xylem vulnerability in three contrasting coffee (*Coffea arabica*) cultivars. Tree Physiology. 2000;**20**:159-168. DOI: 10.1093/tree.phys/20.3.159

[41] Lima ALS, DaMatta FM, Pinheiro HA, Totola MR, Loureiro ME. Photochemical responses and oxidative stress in two clones of *Coffea canephora* under water deficit conditions. Environmetal and Experimental Botany. 2002;**47**:239-247. DOI: https://doi.org/10.1016/S0098-8472(01)00130-7

[42] DaMatta FM, Chaves ARM, Pinheiro HA, Ducatti C, Loureiro ME. Drought tolerance of two field-grown clones of *Coffea canephora*. Plant Science. 2003;**164**:111-117. DOI: https://doi.org/10.1016/S0168-9452(02)00342-4

[43] Pinheiro HA, DaMatta FM, Chaves AR, Loureiro ME, Ducatti C. Drought tolerance is associated with rooting depth and stomatal control of water use in clones of *Coffea canephora*. Annals of Botany. 2005;**96**:101-108. DOI: https://doi.org/10.1093/aob/mci154

[44] Nardini A, Ôunapuu-Pikas E, Savi T. When smaller is better: Leaf hydraulic conductance and drought vulnerability correlate to leaf size and venation density across four *Coffea arabica* genotypes. Functional Plant Biology. 2014;**41**:972-982. DOI: 10.1071/FP13302

[45] Menezes-Silva PE, Cavatte PC, Martins SCV, Reis JV, Pereira LF, Ávila RT, Almeida AL, Ventrella MC, DaMatta FM. Wood density, but not leaf hydraulic architecture, is associated with drought tolerance in clones of *Coffea canephora*. Trees. 2015;**29**:1687-1697. DOI: 10.1007/s00468-015-1249-5

[46] Miniussi M, Terra LD, Savi T, Pallavicini A, Nardini A. Aquaporins in *Coffea arabica* L.: Identification, expression, and impacts on plant water relations and hydraulics. Plant Physiology and Biochemistry. 2015;**95**:92-102. DOI: 10.1016/j.plaphy.2015.07.024

[47] Moffato LS, Carneiro FA, Vieira NG, Duarte KE, Vidal RO, Alekcevetch JC, Cotta MG, Verdeil J-L, Lapeyre-Montes F, Lartad M, Leroy T, Bellis FD, Pot D, Rodrigues GC,

Carazzolle MF, Pereira GAG, Andrade AC, Marraccini P. Identification of candidate genes for drought tolerance in coffee by high-throughput sequencing in the shoot apex of different *Coffea arabica* cultivars. BMC Plant Biology. 2016;**16**:94. DOI: https://doi.org/10.1186/s12870-016-0777-5

[48] Menezes-Silva PE, Sanglard LMPV, Ávila RT, Morais LE, Martins SCV, Nobres P, Patreze CM, Ferreira MA, Araújo WL, Fernie AR, DaMatta FM. Photosynthetic and metabolic acclimation to repeated drought events play key roles in drought tolerance in coffee. Journal of Experimental Botany. 2017;**68**:4309-4322. DOI: 10.1093/jxb/erx211

[49] Martins MQ, Rodrigues WP, DaMatta FM, Ramalho JDC, Partelli FL. O Cafeeiro no contexto de alterações climáticas. In: Ferreira a, Partelli FL, do Amaral JAT , Dalvi LP, Caldeira MVW, Coelho RI. Tópicos Especiais Em Genética e Melhoramento. Alegre: Suprema Gráfica e Editora; 2016. pp. 263-288

[50] ICO (International Coffee Organization). World Coffee Trade (1963–2013): A Review of the Markets, Challenges and Opportunities Facing the Sector. ICC (International Coffee Council) 111-5 Rev. 1. 2014. 29 p. Available at http://www.ico.org/news/icc-111-5-r1e-world-coffee-outlook.pdf

[51] Ramalho JC, DaMatta FM, Rodrigues AP, Scotti-Campos P, Pais I, Batista-Santos P, Partelli FL, Ribeiro A, Lidon FC, Leitão AE. Cold impact and acclimation response of *Coffea* spp. plants. Theoretical and Experimental Plant Physiology. 2014;**26**:5-18. DOI: 10.1007/s40626-014-0001-7

[52] Waller JM, Bigger M, Hillocks RJ. World coffee production. In: Waller J, editor. Coffee Pests, Diseases and Their Management. Chapter 2. Egham, Surrey, UK: CAB International; 2007. pp. 17-40. DOI: 10.1079/9781845931292.0017

[53] Osorio N. The Global Coffee Crisis: A Threat to Sustainable Development. London: International Coffee Organization; 2002. p. 4

[54] ICO (International Coffee Organization). Trade statistics. Available from: http://www.ico.org/historical/1990%20onwards/PDF/1a-total-production.pdf [Accessed: 8 September 2017]

[55] Bagyaraj DJ, Thilagar G, Ravisha C, Kushalappa CG, Krishnamurthy KN, Vaast P. Below ground microbial diversity as influenced by coffee agroforestry systems in the Western Ghats, India. Agriculture, Ecosystems and Environment. 2015;**202**:198-202. DOI: https://doi.org/10.1016/j.agee.2015.01.015

[56] Krishnan S, Ranker TA, Davis AP, Rakotomalala JJ. An assessment of the genetic integrity of ex situ germplasm collections of three endangered species of *Coffea* from Madagascar: Implications for the management of field germplasm collections. Genetic Resources Crop Evolution. 2013;**60**:1021-1036. DOI: 10.1007/s10722-012-9898-3

[57] Partelli FL, Batista-Santos P, Scotti-Campos P, Pais IP, Quartin VL, Vieira HD, Ramalho JC. Characterization of the main lipid components of chloroplast membranes and cold induced changes in *Coffea* spp. Environmental and Experimental Botany. 2011;**74**:194-204. DOI: 10.1016/j.envexpbot.2011.06.001

[58] Davis AP, Tosh J, Ruch N, Fay MF. Growing coffee: *Psilanthus* (Rubiaceae) subsumed on the basis of molecular and morphological data implications for the size, morphology, distribution and evolutionary history of *Coffea*. Botanical Journal of the Linnean Society. 2011;**167**:357-377. DOI: 10.1111/j.1095-8339.2011.01177.x

[59] Farah A, Monteiro MC, Calado V, Franca AS, Trugo LC. Correlation between cup quality and chemical attributes of Brazilian coffee. Food Chemistry. 2006;**98**:373-380. DOI: 10.1016/j.foodchem.2005.07.032

[60] Joët T, Laffargue A, Descroix F, Doulbeau S, Bertrand B, Kochko A, Dussert S. Influence of environmental factors, wet processing and their interactions on the biochemical composition of green Arabica coffee beans. Food Chemistry. 2010;**118**:693-701. DOI: 10.1016/j.foodchem.2009.05.048

[61] Bicho NC, Leitão AE, Ramalho JC, Alvarenga NBD, Lidon FC. Identification of nutritional descriptors of roasting intensity in beverages of Arabica and Robusta coffee beans. International Journal of Food Sciences and Nutrition. 2011;**62**:865-871. DOI: 10.3109/09637486.2011.588594

[62] Bertrand B, Boulanger R, Dussert S, Ribeyre F, Berthiot L, Descroix F, Joët T. Climatic factors directly impact the volatile organic compound fingerprint in green Arabica coffee bean as well as coffee beverage quality. Food Chemistry. 2012;**135**:2575-2583. DOI: 10.1016/j.foodchem.2012.06.060

[63] Ribeiro VS, Leitão AE, Ramalho JC, Lidon FC. Chemical characterization and antioxidant properties of a new coffee blend with cocoa, coffee silverskin and green coffee minimally processed. Food Research International. 2014;**61**:39-47. DOI: http://dx.doi.org/10.1016/j.foodres.2014.05.003

[64] Cheng B, Furtado A, Smyth HE, Henry RJ. Influence of genotype and environment on coffee quality. Trends in Food Science & Technology. 2016;**57**:20-30. DOI: http://dx.doi.org/10.1016/j.tifs.2016.09.003

[65] Camargo AP. O clima e a cafeicultura no Brasil. Informe Agropecuário. 1985;**11**:13-26. ISBN: 01003364

[66] Scotti-Campos P, Pais IP, Partelli FL, Batista-Santos P, Ramalho JC. Phospholipids profile in chloroplasts of *Coffea* spp. genotypes differing in cold acclimation ability. Journal of Plant Physiology. 2014;**171**:243-249. DOI: https://doi.org/10.1016/j.jplph.2013.07.007

[67] Charrier A, Berthaud J. Botanical classification of coffee. In: Clifforf MN, Willson KC, editors. Coffee: Botany, Biochemistry, and Production of Beans and Beverage. London: Crom Helm; 1985. pp. 13-47

[68] Coste R. Coffee: The Plant and the Product. London: MacMillan Press Ltd; 1992. 328 p

[69] Davis AP, Govaerts R, Bridson DM, Stoffelen P. An annotated taxonomic conspectus of the genus *Coffea* (*Rubiaceae*). Botanical Journal of the Linnean Society. 2006;**152**:465-512. DOI: 10.1111/j.1095-8339.2006.00584.x

[70] Partelli FL, Vieira HD, Silva MG, Ramalho JC. Seasonal vegetative growth of different age branches of conilon coffee tree. Semina: Ciências Agrárias. 2010;**31**:619-626. DOI: 10.5433/1679-0359.2010v31n3p619

[71] Partelli FL, Marré WB, Falqueto AR, Vieira HD, Cavatti PC. Seasonal vegetative growth in genotypes of *Coffea canephora*, as related to climatic factors. Journal of Agricultural Science. 2013;**5**:108-116. DOI: http://dx.doi.org/10.5539/jas.v5n8p108

[72] Covre AM, Partelli FL, Bonomo R, Braun H, Ronchi CP. Vegetative growth of Conilon coffee plants under two water conditions in the Atlantic region of Bahia State, Brazil. Acta Scientiarum. Agronomy. 2016;**38**:535-545. DOI: http://dx.doi.org/10.4025/actasciagron.v38i4.30627

[73] Dubberstein D, Partelli FL, Dias JRM, Espindula MC. Influence of fertilization on coffee trees vegetative growth in west south Amazon. Coffee Science. 2017;**12**:197-206. Available at: http://www.coffeescience.ufla.br/index.php/Coffeescience/article/view/1228

[74] Taques RC, Dadalto GG. Zoneamento agroclimático para a cultura do café Conilon no Estado do Espírito Santo. In: Ferrão RG, Fonseca AFA, Ferrão MAG, De Muner LH. Café Conilon. 2nd ed. Vitória: Incaper; 2017. pp. 69-79

[75] Xu Z, Jiang Y, Zhou G. Response and adaptation of photosynthesis, respiration, and antioxidant systems to elevated CO_2 with environmental stress in plants. Frontiers in Plant Science. 2015;**6**:701-717. DOI: https://doi.org/10.3389/fpls.2015.00701

[76] Jiang Y, Xu Z, Zhou G, Liu T. Elevated CO_2 can modify the response to a water status gradient in a steppe grass: From cell organelles to photosynthetic capacity to plant growth. BMC Plant Biology. 2016;**16**:157. DOI: 10.1186/s12870-016-0846-9

[77] Song Y, Yu J, Huang B. Elevated CO_2-mitigation of high temperature stress associated with maintenance of positive carbon balance and carbohydrate accumulation in Kentucky bluegrass. PLoS One. 2014;**9**:e89725. DOI: https://doi.org/10.1371/journal.pone.0089725

[78] Djanaguiraman M, Vara Prasad PV. Ethylene production under high temperature stress causes premature leaf senescence in soybean. Functional Plant Biology. 2010;**37**:1071-1084. DOI: 10.1071/FP10089

[79] Wise RR, Olson AJ, Schrader SM, Sharkey TD. Electron transport is the functional limitation of photosynthesis in field-grown pima cotton plants at high temperature. Plant Cell & Environment. 2004;**27**:717-724. DOI: 10.1111/j.1365-3040.2004.01171.x

[80] Wahid A, Gelani S, Ashraf M, Foolad MR. Heat tolerance in plants: An overview. Environmental and Experimental Botany. 2007;**61**:199-223. DOI: 10.1016/j.envexpbot.2007.05.011

[81] Lambers H, Chapin III FS, Pons TL. Plant Physiological Ecology. 2nd ed. New York: Springer-Verlag; 2008. 604 p. DOI: 10.1007/978-0-387-78341-3

[82] Crafts-Brandner SJ, Salvucci ME. Rubisco activase constrains the photosynthetic potential of leaves at high temperature and CO_2. Proceedings of the National Academy of Sciences of the United States of America. 2000;**97**:13430-13435. DOI: 10.1073/pnas.230451497

[83] Ainsworth EA, Rogers A. The response of photosynthesis and stomatal conductance to rising [CO_2]: Mechanisms and environmental interactions. Plant Cell & Environment. 2007;**30**:258-270. DOI: 10.1111/j.1365-3040.2007.01641.x

[84] Barnabás B, Jäger K, Fehér A. The effect of drought and heat stress on reproductive processes in cereals. Plant Cell Environment. 2008;**31**:11-38. DOI: 10.1111/j.1365-3040.2007.01727.x

[85] Assad ED, Pinto HS, Zullo Jr J, Ávila AMH. Impacto das mudanças climáticas no zoneamento agroclimático do café no Brasil. Pesquisa Agropecuária Brasileira. 2004;**39**:1057-1064. DOI: http://dx.doi.org/10.1590/S0100-204X2004001100001

[86] Martins MQ, Fortunato AS, Rodrigues WP, Partelli FL, Campostrini E, Lidon FC, DaMatta FM, Ramalho JC, Ribeiro-Barros AI. Selection and validation of reference genes for accurate RT-qPCR data normalization in *Coffea* spp. under a climate changes context of interacting elevated [CO_2] and temperature. Frontier. Plant Science. 2017;**8**:307. DOI: https://doi.org/10.3389/fpls.2017.00307

[87] Barros R, Mota JWS, DaMatta FM, Maestri M. Decline of vegetative growth in *Coffea arabica* L. in relation to leaf temperature: Water potential and stomatal conductance. Field Crops Research. 1997;**54**:65-72. DOI: https://doi.org/10.1016/S0378-4290(97)00045-2

[88] Rodrigues WP, Machado Filho JA, Silva JR, Figueiredo FAMMA, Ferraz TM, Ferreira LS, Bezerra LS, Abreu DP, Bernado WP, Cespom L, Sousa EF, Glenn DM, Ramalho JC, Campostrini E. Whole-canopy gas exchange in *Coffea* sp. is affected by supra-optimal temperature and light distribution within the canopy: The insights from an improved multi-chamber system. Scientia Horticulturae. 2016;**211**:194-202. DOI: https://doi.org/10.1016/j.scienta.2016.08.022

[89] Thioune EH, McCarthy J, Gallagher T, Osborne BA. Humidity shock leads to rapid, temperature dependent changes in coffee leaf physiology and gene expression. Tree Physiology. 2017;**37**:367-379. DOI: 10.1093/treephys/tpw129

[90] Noctor G, Mhamdi A, Foyer CH. The roles of reactive oxygen metabolism in drought: Not so cut and dried. Plant Physiology. 2014;**164**:1636-1648. DOI: 10.1104/pp.113.233478

[91] Barbosa MR, Silva MMDA, Willadino L, Ulisses C, Camara TR. Geração e desintoxicação enzimática de espécies reativas de oxigênio em plantas. Ciência Rural. 2014;**44**:453-460. DOI: http://dx.doi.org/10.1590/S0103-84782014000300011

[92] Pinheiro HA, DaMatta FM, Chaves ARM, Loureiro ME, Ducatti C. Drought tolerance in relation to protection against oxidative stress in clones of *Coffea canephora* subjected to long-term drought. Plant Science. 2004;**167**:1307-1314. DOI: https://doi.org/10.1016/j.plantsci.2004.06.027

[93] Martins SCV, Araújo WA, Tohge T, Fernie AR, DaMatta FM. High-light-acclimated coffee plants the metabolic machinery is adjusted to avoid oxidative stress rather than to benefit from extra light enhancement in photosynthetic yield. PLoS One. 2014;**9**: e94862. DOI: https://doi.org/10.1371/journal.pone.0094862

[94] Trueba S, Isnard S, Barthélémy D, Olson ME. Trait coordination, mechanical behaviour and growth form plasticity of *Amborella trichopoda* under variation in canopy openness. AoB Plants. 2016;**8**:plw068. DOI: 10.1093/aobpla/plw068

[95] Meinzer FC, Goldstein G, Grantz DA. Carbon isotope discrimination in coffee genotypes grown under limited water supply. Plant Physiology. 1990;**92**:130-135. DOI: 0032-0889/90/92/0130/06/$01.00/0

[96] Zhu C, Ziska L, Zhu J, Zeng Q, Xie Z, Tang H, Jia X, Hasegawa T. The temporal and species dynamics of photosynthetic acclimation in flag leaves of rice (*Oryza sativa*) and wheat (*Triticum aestivum*) under elevated carbon dioxide. Physiologia Plantarum. 2012;**145**:395-405. DOI: 10.1111/j.1399-3054.2012.01581.x

[97] Long SP, Ainsworth EA, Rogers A, Ort DR. Rising atmospheric carbon dioxide: Plants FACE the future. Annual Review Plant Biology. 2004;**55**:591-628. DOI: 10.1146/annurev.arplant.55.031903.141610

[98] Rakocevic M, Ferrandes R, Marchiori PER, Ribeiro RV. Estimating the canopy architecture and photosynthesis of *Coffea arabica* L. plants cultivated under long-term elevated air CO_2 concentration. In: International Conference on Functional-Structural Plant Growth Modeling, Simulation, Visualization and Applications (FSPMA); 7–11 Nov 2016; Qingdao. New York: IEEE; 2017. pp. 175-182. DOI: 10.1109/FSPMA.2016.7818304

[99] Wertin TM, McGuire MA, Teskey RO. The influence of elevated temperature, elevated atmospheric CO_2 concentration and water stress on net photosynthesis of loblolly pine (*Pinus taeda* L.) at northern, central and southern sites in its native range. Global Change Biology. 2010;**16**:2089-2103. DOI: 10.1111/j.1365-2486.2009.02053.x

[100] Nunes MA, Ramalho JC, Dias MA. Effect of nitrogen supply on the photosynthetic performance of leaves from coffee plants exposed to bright light. Journal of Experimental Botany. 1993;**44**:893-899. DOI: 10.1093/jxb/44.5.893

[101] Ramalho JC, Campos PS, Teixeira M, Nunes MA. Nitrogen dependent changes in antioxidant systems and in fatty acid composition of chloroplast membranes from *Coffea arabica* L. plants submitted to high irradiance. Plant Science. 1998;**135**:115-124. DOI: https://doi.org/10.1016/S0168-9452(98)00073-9

[102] Ramalho JC, Pons T, Groeneveld H, Azinheira HG, Nunes MA. Photosynthetic acclimation to high light conditions in mature leaves of *Coffea arabica* L.: Role of xanthophylls, quenching mechanisms and nitrogen nutrition. Functional Plant Biology. 2000;**27**:43-51. DOI: 10.1071/PP99013

[103] Ramalho JC, Fortunato AS, Goulao LF, Lidon FC. Cold-induced changes in mineral content in *Coffea* spp. leaves – identification of descriptors for tolerance assessment. Biologia Plantarum. 2013;**57**:495-506. DOI: 10.1007/s10535-013-0329-x

[104] Ramalho JC, Rebelo MC, Santos ME, Antunes ML, Nunes MA. Effects of calcium deficiency on *Coffea arabica*. Nutrient changes and correlation of calcium levels with some photosynthetic parameters. Plant and Soil. 1995;**172**:87-96. DOI: 10.1007/BF00020862

[105] Blank RR, Morgan T, Ziska LH, White RH. Effect of atmospheric CO_2 levels on nutrients in cheatgrass tissue. Natural Resources and Environmental Issues. 2011;**16**:1-6. Available form: http://digitalcommons.usu.edu/nrei/vol16/iss1/18

[106] Chaturvedi AK, Bahuguna RN, Pal M, Shah D, Maurya S, Jagadish KSV. Elevated CO_2 and heat stress interactions affect grain yield, qualityand mineral nutrient composition in rice under field conditions. Field Crops Research. 2017;**206**:149-157. DOI: http://dx.doi.org/10.1016/j.fcr.2017.02.018

[107] Conroy J, Hocking P. Nitrogen nutrition of C3 plants at elevated atmospheric CO_2 concentrations. Physiologia Plantarum. 1993;**89**:570-576. DOI: 10.1111/j.1399-3054.1993.tb05215.x

[108] Thiec DL, Dixon M, Loosveldt P, Garrec JP. Seasonal and annual variations of phosphorus, calcium, potassium and manganese contents in different cross-sections of *Picea abies* (L.) karst. Needles and *Quercus rubra* L. leaves exposed to elevated CO_2. Trees. 1995;**10**:55-62. DOI: https://doi.org/10.1007/BF00192184

[109] Cotrufo MF, Ineson P, Scott A. Elevated CO_2 reduces the nitrogen concentration of plant tissues. Global Change Biology. 1998;**4**:43-54. DOI: 10.1046/j.1365-2486.1998.00101.x

[110] Overdieck D. Elevated CO_2 and the mineral content of herbaceous and woody plants. Vegetatio. 1993;**104**:403-411. DOI: https://doi.org/10.1007/BF00048169

[111] Manderscheid R, Bender J, Jäger H-J, Weigel HJ. Effects of season long CO_2 enrichment on cereals. II. Nutrient concentrations and grain quality. Agriculture, Ecosystems & Environment. 1995;**54**:175-185. DOI: https://doi.org/10.1016/0167-8809(95)00602-O

[112] Fangmeier A, Grüters U, Högy P, Vermehren B, Jäger HJ. Effects of elevated CO_2, nitrogen supply, and tropospheric ozone on spring wheat. II. Nutrients (N, P, K, S, Ca, mg, Fe, Mn, Zn). Environmental Pollution. 1997;**96**:43-59. DOI: 10.1016/S0269-7491(97)00013-4

[113] Leakey ADB, Ainsworth EA, Bernacchi CJ, Rogers A, Long SP, Ort DR. Elevated CO_2 effects on plant carbon, nitrogen, and water relations: Six important lessons from FACE. Journal of Experimental Botany. 2009;**60**:2859-2876. DOI: https://doi.org/10.1093/jxb/erp096

[114] Fortunato A, Lidon FC, Batista-Santos P, Leitão AE, Pais IP, Ribeiro AI, Ramalho JC. Biochemical and molecular characterization of the antioxidative system of *Coffea* sp. under cold conditions in genotypes with contrasting tolerance. Journal of Plant Physiology. 2010;**167**:333-342. DOI: 10.1016/j.jplph.2009.10.013

[115] DaMatta FM, Ronchi CP, Maestri M, Barros RS. Ecophysiology of coffee growth and production. Brazilian Journal of Plant Physiology. 2008;**19**:485-510. DOI: http://dx.doi.org/10.1590/S1677-04202007000400014

[116] Kumar D, Tieszen LL. Photosynthesis in *Coffea arabica*. L. Effects of light and temperature. Experimental Agriculture. 1980;**16**:13-19. DOI: 10.1017/S0014479700010656

[117] Rodríguez-López CPC, Menezes-Silva PE, Martins SCV, Morais LE, Medina EF, DaMatta FM. Physiological and biochemical abilities of robusta coffee leaves for acclimation to cope with temporal changesin light availability. Physiologia Plantarum. 2013;**149**:45-55. DOI: 10.1111/ppl.12010

[118] Cavatte PC, Oliveira AA, Morais LE, Martins SC, Sanglard LM, DaMatta FM. Could shading reduce the negative impacts of drought on coffee? A morphophysiological analysis. Physiologia Plantarum. 2012;**114**:111-122. DOI: 10.1111/j.1399-3054.2011.01525.x

[119] Beer J. Advantages, disadvantages and desirable characteristics of shade trees for coffee, cocoa and tea. Agroforesty Systemys. 1987;**5**:3-13. DOI: https://doi.org/10.1007/BF000 46410

[120] Souza HN, Graaff J, Pullemam MM. Strategies and economics of farming systems with coffee in the Atlantic rainforest biome. Agroforestry Systems. 2012;**84**:227-242. DOI: https://doi.org/10.1007/s10457-011-9452-x

[121] Barradas VL, Fanjul L. Microclimatic chacterization of shaded and open-grown coffee (*Coffea arabica* L.) plantations in Mexico. Agricultural and Foresty Metereology. 1986;**38**:101-112. DOI: https://doi.org/10.1016/0168-1923(86)90052-3

[122] Vaast P, Angrand J, Franck N, Dauzat J, Genard M. Fruit load and branch ring-barking affect carbon allocation and photosynthesis of leaf and fruit of *Coffea arabica* in the field. Tree Physiology. 2005;**25**:753-760. DOI: 10.1093/treephys/25.6.753

[123] Siebert SF. From shade- to sun-grown perennial crops in Sulawesi, Indonesia: Implications for biodiversity conservation and soil fertility. Biodiversity and Conservation. 2002;**11**:1889-1902. DOI: https://doi.org/10.1023/A:1020804611740

[124] Scherr SJ, McNeely JA. Biodiversity conservation and agricultural sustainability: Towards a new paradigm of 'ecoagriculture' landscapes. Philosophical Transactions of the Royal Society. 2008;**363**:477-494. DOI: 10.1098/rstb.2007.2165

[125] Siles P, Harmand JM, Vaast P. Effects of *Inga densiflora* on the microclimate of coffee (*Coffea arabica* L.) and overall biomass under optimal growing conditions in Costa Rica. Agroforestry Systems. 2010;**78**:269-286. DOI: 10.1007/s10457-009-9241-y

[126] Charbonnier F, Roupsard O, Maire GL, Guillemot J, Casanoves F, Lacointe A, Vaast P, Allinne C, Audebert L, Cambou A, Clément-Vidal A, Defrenet E, Duursma RA, Jarri L, Jourdan C, Khac E, Leandro P, Medlyn BE, Saint-André L, Thaler P, Meersche KVD, Aguilar AB, Lehner P, Dreyer E. Increased light-use efficiency sustains net primary productivity of shaded coffee plants in agroforestry system. Plant, Cell and Environment. 2017;**40**:1592-1608. DOI: 10.1111/pce.12964

[127] Bragança SM, Carvalho CHS, Fonseca AFA, Ferrão RG. Clonal varieties of Conilon coffee for the Espírito Santo state, Brazil. Pesquisa Agropecuária Brasileira. 2001;**36**: 765-770. DOI: http://dx.doi.org/10.1590/S0100-204X2001000500006

[128] Oliosi G, Giles JD, Rodrigues WP, Ramalho JC, Partelli LP. Microclimate and development of *Coffea canephora* cv. Conilon under different shading levels promoted by

Australian cedar (*Toona ciliate* M. Roem. Var. Australis). Australian Journal of Crop Science. 2016;**10**:528-538. DOI: 10.21475/ajcs.2016.10.04.p7295x

[129] Partelli FL, Araújo AV, Vieira HD, Dias JRM, Menezes LFT, Ramalho JC. Microclimate and development of 'Conilon' coffee intercropped with rubber trees. Pesquisa Agropecuária Brasileira. 2014;**49**:872-881. DOI: http://dx.doi.org/10.1590/S0100-204X2014001100006

[130] Ricci MSF, Costa JR, Pinto AN, Santos VLS. Organic cultivation of coffee cultivars grown under full sun and under shading. Pesquisa Agropecuária Brasileira. 2006;**41**:569-575. DOI: http://dx.doi.org/10.1590/S0100-204X2006000400004

[131] Somporn C, Kamtuo A, Theerakulpisut P, Siriamornpun S. Effect of shading on yield, sugar content, phenolic acids and antioxidant property of coffee beans (*Coffea arabica* L. cv. Catimor) harvested from North-Eastern Thailand. Journal of the Science of Food and Agriculture. 2012;**92**:1956-1963. DOI: 10.1002/jsfa.5568

[132] Muschler RG. Shade improves coffee quality in a sub-optimal coffee zone of Costa Rica. Agroforesty Systems. 2001;**85**:131-139. DOI: http://dx.doi.org/10.1023/A:1011863426305

[133] Vaast P, Bertrand B, Perriot JJ, Guyot B, Genard M. Fruit thinning and shade improve bean characteristics and beverage quality of coffee (*Coffea arabica* L.) under optimal conditions. Journal of the Science of Food and Agriculture. 2006;**86**:197-204. DOI: 10.1002/jsfa.2338

[134] Camargo APD, Camargo MBD. Definition and outline for the phenological phases of arabica coffee under Brazilian tropical conditions. Bragantia. 2001;**60**:65-68. DOI: http://dx.doi.org/10.1590/S0006-87052001000100008

[135] Glenn DM, Prado E, Erez A, McFerson J, Puterka GJA. Reflective, processed kaolin particle film affects fruit temperature, radiation reflection, and solar injury in apple. Journal of the American Society for Horticultural Science. 2002;**127**:188-193. Available from: https://www.ars.usda.gov/ARSUserFiles/2017/Sunburn%20paper%20ASHS.pdf

[136] Glenn DM. Particle film mechanisms of action that reduce the effect of environmental stress in "Empire" apple. Journal of the American Society for Horticultural Science. 2009;**134**:314-321. Available at: http://journal.ashspublications.org/content/134/3/314.full

[137] Glenn DM, Cooley N, Shellie K. Impact of kaolin particle film and water deficit on wine grape water use efficiency and plant water relations. Hortscience. 2010;**45**:1178-1187. Available at: http://hortsci.ashspublications.org/content/45/8/1178.abstract

[138] Brilhante L, Belfiore N, Gaiotti F, Lovat L, Sansone L, Poni S, Tomasi D. Comparing kaolin and pinolene to improve sustainable grapevine production during drought. PLoS One. 2016;**110**:e0156631. DOI: 10.1371/journal.pone.0156631

[139] Steiman SR, Bittenbender HB. Analysis of kaolin particle film use and its application on coffee. Hortscience. 2007;**42**:1605-1608. Available from: http://hortsci.ashspublications.org/content/42/7/1605.full

[140] Verdin-Filho AC, Tomaz MA, Ferrão RG, Ferrão MAG, Fonseca AFA, Rodrigues WN. Conilon coffee yield using the programmed pruning cycle different cultivation densities.

Coffee Science. 2014;**9**:489-494. Available at: http://www.coffeescience.ufla.br/index.php/Coffeescience/article/download/734/pdf_130

[141] Zuazo VHD, Pleguezuelo CRR. Soil-erosion and runoff prevention by plant covers. A review. Agronomy for Sustainable Development. 2008;**28**:65-86. DOI: 10.1051/agro:2007062

[142] Santana AO, Cuniat G, Imaña-Encinas J. Contribution of understory vegetation on total evapotranspiration in distinct environments of the cerrado biome, Distrito Federal, Brazil. Ciência Florestal. 2010;**20**:269-281. DOI: http://dx.doi.org/10.5902/198050981851

[143] Alcântara EN, Ferreira MM. Effects of weed control methods in coffee (*Coffea arabica* L.) on soil physical quality. Revista Brasileira de Ciência do Solo. 2000;**24**:711-721. DOI: http://dx.doi.org/10.1590/S0100-06832000000400003

[144] Alcântara EN de, Nobrega JCA, Ferreira MM. Métodos de controle de plantas daninhas no cafeeiro afetam os atributos químicos do solo. Ciência Rural. 2009;**39**:749-757. DOI: http://dx.doi.org/10.1590/S0103-84782009000300018

[145] Effgen TAM, Passos RR, Andrade FV, Lima JSS, Reis EF, Borges EN. Physical soil properties as a function of management in crops of conilon coffee. Revista Ceres. 2012;**59**:414-421. DOI: 10.1590/S0034-737X2012000300018

[146] Fernandes ALT, Santinato F, Ferreira RT, Santinado R. Reduction of mineral fertilization of arabic coffee using coffee straw. Coffee Science. 2013;**8**:324-336. Available from: http://www.coffeescience.ufla.br/index.php/Coffeescience/article/view/454/pdf_44

[147] Paes JMV, Andreola F, Brito CH, Loures EG, Decomposição d. Palha de café em três tipos de solo e sua influência sobre a CTC e o pH. Revista Ceres. 1996;**43**:674-683. Available from: http://www.ceres.ufv.br/ojs/index.php/ceres/article/viewFile/2373/381

[148] Almeida FS. Efeitos alelopáticos de resíduos vegetais. Pesquisa Agropecuária Brasileira. 1991;**26**:221-236. Available from: https://seer.sct.embrapa.br/index.php/pab/article/view/3333/666

[149] Sharma E, Rai SC, Sharma R. Soil, water and nutrient conservation in mountain farming systems: Case-study from the Sikkim Himalaya. Journal of Environmental Management. 2001;**61**:123-135. DOI: 10.1006/jema.2000.0386

[150] Hammad AA, Haugen LE, Borresen T. Effects of stonewalled terracing techniques on soil-water conservation and wheat production under mediterranean conditions. Environmental Management. 2004;**34**:701-710. DOI: 10.1007/s00267-003-0278-9

[151] Al-Seekh SH, Mohammad AG. The effect of water harvesting techniques on runoff, sedimentation, and soil properties. Environmental Management. 2009;**44**:37-45. DOI: 10.1007/s00267-009-9310-z

[152] Magalhães GMF. Analysis of efficiency of retention terraces in sub-basins of the São Francisco River. Revista Brasileira de Engenharia Agrícola e Ambiental. 2013;**17**:1109-1115. DOI: http://dx.doi.org/10.1590/S1415-43662013001000013

Section 2

Climate Resilient Strategies

Chapter 5

Agricultural Management Impact on Greenhouse Gas Emissions

Upendra M. Sainju

Additional information is available at the end of the chapter

http://dx.doi.org/10.5772/intechopen.72368

Abstract

Management practices used on croplands to enhance crop yields and quality can contribute about 10–20% of global greenhouse gases (GHGs: carbon dioxide [CO_2], nitrous oxide [N_2O], and methane [CH_4]). Some of these practices are tillage, cropping systems, N fertilization, organic fertilizer application, cover cropping, fallowing, liming, etc. The impact of these practices on GHGs in radiative forcing in the earth's atmosphere is quantitatively estimated by calculating net global warming potential (GWP) which accounts for all sources and sinks of CO_2 equivalents from farm operations, chemical inputs, soil carbon sequestration, and N_2O and CH_4 emissions. Net GWP for a crop production system is expressed as kg CO_2 eq. ha^{-1} $year.^{-1}$ Net GWP can also be expressed in terms of crop yield (kg CO_2 eq. kg^{-1} grain or biomass yield) which is referred to as net greenhouse gas intensity (GHGI) or yield-scaled GWP and is calculated by dividing net GWP by crop yield. This article discusses the literature review of the effects of various management practices on GWP and GHGI from croplands as well as different methods used to calculate net GWP and GHGI. The paper also discusses novel management techniques to mitigate net CO_2 emissions from croplands to the atmosphere. This information will be used to address the state of global carbon cycle.

Keywords: crop yield, greenhouse gas, global warming, potential, management practice, soil carbon sequestration

1. Overview

Management practices on croplands can contribute about 10–20% of global greenhouse gases (GHGs: carbon dioxide [CO_2], nitrous oxide [N_2O], and methane [CH_4]) [1, 2]. Quantitative estimate of the impact of these GHGs in radiative forcing in the earth's atmosphere is done by calculating net global warming potential (GWP) which accounts for all sources and sinks of CO_2 equivalents from farm operations, chemical inputs, soil carbon (C) sequestration, and N_2O and

CH_4 emissions [3, 4]. Net GWP for a crop production system is expressed as kg CO_2 eq. ha^{-1} $year.^{-1}$ Net GWP can also be expressed in terms of crop yield (kg CO_2 eq. kg^{-1} grain or biomass yield) which is referred to as net greenhouse gas intensity (GHGI) or yield-scaled GWP and is calculated by dividing net GWP by crop yield [3]. These values can be affected both by net GHG emissions and crop yields. Sources of GHGs in agroecosystems include N_2O and CH_4 emissions (or CH_4 uptake) as well as CO_2 emissions associated with farm machinery used for tillage, planting, harvesting, and manufacture, transportation, and applications of chemical inputs, such as fertilizers, herbicides, and pesticides, while soil carbon sequestration rate can be either a sink or source of CO_2 [4–6]. In the calculations of net GWP and GHGI, emissions of N_2O and CH_4 are converted into their CO_2 equivalents of global warming potentials which are 265 and 28, respectively, for a time horizon of 100 year [2]. The balance between soil carbon sequestration rate, N_2O and CH_4 emissions (or CH_4 uptake), and crop yield typically controls net GWP and GHGI [3, 4, 7].

Some of the improved management practices used for reducing net GWP and GHGI from croplands are no-till, increased cropping intensity, diversified crop rotation, cover cropping, and reduced N fertilization rates [3, 4, 7–10]. Soil organic carbon can usually be increased by adopting no-tillage practice which decreases microbial activity and CO_2 emissions as a result of reduced soil disturbance and residue incorporation compared with conventional tillage practice [3, 11]. Tillage, however, can interact with crop residue carbon input on soil carbon sequestration which varies by region [12]. When carbon input is <15% due to reduced crop yields, soil organic carbon is often lower with no-tillage than conventional tillage in regions with wetter and cooler climate, such as in eastern USA and Canada [12]. In such regions, tillage often redistributes crop residues in the soil profile, resulting in lower soil organic carbon in the surface soil and greater in the subsurface soil, with overall lower soil profile carbon is greater in conventional tillage than no-tillage [13, 14]. The reverse is true when carbon input is >15% [12]. With double rather than single crop in a year, Luo et al. [14] found that no-tillage increased soil organic carbon in the soil profile compared with conventional tillage. In regions with subtropical humid and semiarid climates, such as in southern USA and northern Great Plains, no-tillage can increase soil organic carbon compared with conventional tillage by increasing carbon input as well as reducing soil organic matter mineralization [12].

Soil organic carbon can also be increased by increasing the quality and quantity of crop residue returned to the soil due to diversified cropping systems, such as intensive cropping, crop rotation, and cover cropping, compared with non-diversified systems, such as crop-fallow, monocropping, and no cover crop [3, 15]. Crop rotation can increase soil organic carbon by increasing carbon input through increased crop yield compared with monocropping [3, 15, 16]. Similarly, cover cropping can increasing soil organic carbon by increasing the amount of crop residue returned to the soil [17]. In contrast, fallowing can reduce soil organic carbon by reducing carbon input and by increasing soil organic matter mineralization as a result of increased soil temperature and water content [16, 18]. The effect of nitrogen fertilization on soil organic carbon is variable [18–20].

Nitrogen is usually required in large amounts to sustain crop yield and quality compared with other nutrients, such as phosphorus and potassium. Nitrogen fertilization typically stimulates N_2O emissions when the amount of applied nitrogen exceeds crop nitrogen demand [3, 8–10, 21]. Nitrogen fertilization, however, can have a variable effect on CO_2 and CH_4 emissions [15, 22]. Because N_2O emissions has a large effect on net GWP and GHGI, practices that can reduce N fertilization rates without influencing crop yields can substantially reduce net GHG emissions [3, 4]. Other factors that can influence N_2O emissions are type, placement, and method of application of nitrogen fertilizers. Applying nitrogen fertilizer in the spring compared with autumn and using split application compared with one single application at planting can reduce N_2O emissions in some cases [23–25]. Applying N fertilizer at various depths can have variable effects on N_2O emissions [26–29]. Anhydrous ammonia can increase N_2O emissions compared with urea [27, 30, 31]. Similarly, chemical additives to reduce nitrification from nitrogen fertilizers, such as polymer-coated urea and nitrification inhibitors, can substantially reduce N_2O emissions compared with ordinary urea and non-nitrification inhibitors fertilizers [32–34]. Some nitrogen fertilizers, such as urea, emit both CO_2 and N_2O. Nitrogen fertilizers also indirectly emit N_2O through ammonia volatilization and nitrate leaching [31].

Some management practices used for reducing GHG emissions can have adverse effects. Examples of such practices are no-tillage systems where denitrification resulting from higher soil water content can increase N_2O emissions compared with conventional tillage systems in humid regions, thereby reducing the GHG mitigation potential [35]. In contrast, N_2O emissions can be similar [36] or lower [3, 37] with no-tillage compared with conventional tillage in semiarid and arid regions. Sainju et al. [37] reported that crop rotation had no effect on N_2O emissions, but CH_4 uptake was greater with barley-pea rotation than continuous barley in the semiarid region. Cover crops also have variable effect on N_2O emissions [38]. Legume cover crops can increase N_2O emissions compared with nonlegume cover crops during their growth, but emissions can be similar among cover crops when measured over the entire year [38]. Similarly, root respiration and mineralization of crop residue and soil organic carbon can have negative impacts on GHG mitigation, although greater root biomass and distribution can increase carbon sequestration [15, 39]. Therefore, while calculating net GWP and GHGI, all of these factors should be accounted for, regardless of management practices used [3, 4, 40].

Several methods have been used to calculate net GWP and GHGI. Some have used the sum of CO_2 equivalents of N_2O and CH_4 emissions [21, 41, 42], while others [43, 44] have included CO_2 equivalents of all three GHGs. Still others have used CO_2 equivalents of N_2O and CH_4 emissions and soil carbon sequestration rate [45–47]. A full accounting of all sources and sinks of CO_2 emissions to calculate net GWP and GHGI includes CO_2 equivalents from farm operations, N fertilization, and other inputs in addition to above parameters [3, 7, 9, 10, 40, 48–51]. Several researchers have used DAYCENT and GREET models to estimate GWP and GHGI [52, 53]. Some have excluded N_2O and CH_4 emissions, but used CO_2 equivalents of all other sources and sinks [6]. An alternative method of calculating net GWP and GHGI includes substituting soil carbon sequestration rate by soil respiration and the amount of previous year's crop residue returned to the soil [3, 9, 10, 50, 51, 54]. Each method has its own advantages and drawbacks.

2. Impact of management practices

2.1. Tillage

Various studies have shown that both net GWP and GHGI were lower with no-tillage than conventional tillage, regardless of soil and climatic conditions, cropping systems, and methods of calculations [3, 7, 44, 47, 49, 55]; Sainju [56] observed that reductions in net GWP and GHGI due to no-tillage vs. conventional tillage vary among regions with various soil and climatic conditions, but largest difference occurred in sandy soil under moderate annual precipitation (900 mm). Net GWP values, however, increased in regions with higher air temperature. A meta-analysis of nine experiments by the same author on the effect of tillage showed that no-tillage reduced net GWP by 55% and net GHGI by 58% compared with conventional tillage when all sources and sinks of CO_2 were accounted for. With the partial accounting of sources and sinks, the reductions in net GWP and GHGI due to no-tillage vs. conventional tillage were 81 and 73%, respectively, indicating that partial accounting can inflate net GWP and GHGI values [56]. Differences in crop yields among cropping systems and regions resulted in different proportion of reductions in net GWP and GHGI due to no-tillage vs. conventional tillage [56]. Increased soil carbon sequestration rate due to reduced soil disturbance and carbon mineralization reduces net GWP and GHGI in no-tillage [4, 40, 57]. In contrast, increased crop residue incorporation and aeration increases microbial activity which reduces carbon sequestration, thereby reducing net GWP and GHGI in conventional tillage [3, 7, 9]. Reduction in tillage intensity can also reduce net GWP and GHGI [58].

The duration of study can also have a profound influence on net GWP and GHGI with no-tillage vs. conventional tillage. Under corn-soybean rotation in clay loam soil in Colorado, Mosier et al. [3, 7] found that net GWP with no-tillage vs. conventional tillage was lower after 1 year than after 3 year due to differences in soil carbon sequestration rates. In contrast, Six et al. [57] reported that reduction in net GWP with no-tillage vs. conventional tillage was realized only after 10 year in the humid region and 20 year in the dry region due to increased soil aggregation, reduced aeration, and increased soil carbon sequestration. In a meta-analysis of nine experiments, Sainju [56] found that changes in net GWP and GHGI due to no-tillage vs. conventional tillage increased with the duration of the experiment, regardless of the method used for calculation. When soil and climatic conditions, such as soil texture, annual precipitation, and average air temperature of the experimental sites were included in the multiple linear regressions, the relationships were further improved. While air temperature had a negative effect on net GWP and GHGI, the effect of soil texture varied. This could be explained by several factors: (1) no-till can some time increases N_2O emissions due to increased soil water content and denitrification compared with conventional till, especially in the humid region, thereby reducing net GWP and GHGI [4, 40, 57], (2) the potential for soil carbon sequestration using no-tillage decreases and reaches a steady state as the duration of the experiment increases [57, 59], and (3) there is a high uncertainty in spatial and temporal variability in GHG emissions within and among regions due to variations in soil and climatic conditions and management practices [7, 9, 10, 40]. Nevertheless, more long-term experiments are needed to relate the effect of tillage with the duration of experiment on net GWP and GHGI.

2.2. Cropping system

Crop type and cropping systems can affect net GWP and GHGI. Various researchers [3, 7, 48, 49] reported that both net GWP and GHGI were lower with continuous corn than corn-soybean rotation, but net GHGI was lower with soybean than corn when grown alone [53].

Increased soil carbon sequestration due to greater amount of crop residue returned to the soil reduced net GWP and GHGI under continuous corn compared with corn-soybean rotation, although nitrogen fertilization rate to produce sustainable yield was higher in continuous corn [3, 7, 48, 49]. In contrast, greater N_2O emissions following soybean increased net GWP and GHGI in corn-soybean rotation [3, 7, 48, 49]. Under small grain crops, however, several researchers [9, 10, 60, 61] have found that including legumes, such as pea and lentil, in rotation with nonlegumes, such as wheat and barley, reduced net GWP and GHGI compared with continuous nonlegumes. They observed this because (1) no nitrogen fertilizer was applied to legumes compared with nonlegumes which required large amount of nitrogen fertilizers to sustain yields, as nitrogen fertilizer stimulates N_2O emissions and (2) legumes supplied greater amount of nitrogen to succeeding crops due to higher nitrogen concentration when above- and belowground residues were returned to the soil and reduced nitrogen fertilization rate compared with nonlegumes. Sainju et al. [9, 10] also found that legume-nonlegume rotation increased soil carbon sequestration because of increased turnover rate of plant carbon to soil carbon compared to continuous nonlegume.

In a meta-analysis of 11 experiments on the effect of crop rotation containing small and large grain crops on net GWP and GHGI, Sainju [56] reported that crop rotation increased net GWP by 46% and net GHGI by 41% compared with monocropping. This was especially true for large grain crops, such as corn and soybean where net GWP and GHGI were 215 and 325%, respectively, greater under corn-soybean than continuous corn. In contrast, for small grain crops, such as barley and pea, net GWP was 22% lower under barley-pea than continuous barley. Both net GWP and GHGI were 168 and 215%, respectively, lower with perennial than annual crops. Greater number of experiments and magnitude of changes, however, resulted in higher net GWP and GHGI in monocropping than crop rotation under large than small grain crops when values were averaged across experiments during data analysis [56].

As cropping intensity increased, net GWP and GHGI reduced [56]. Greater amount of crop residue returned to the soil and increased carbon sequestration reduced net GWP and GHGI when cropping intensity was increased [9, 50]. Increased carbon sequestration with increased cropping intensity in the semiarid regions with limited precipitation is well known [18, 62]. Several researchers [7, 9, 50] have found that fallowing or crop-fallow rotation increased GHG emissions and therefore net GWP and GHGI compared with continuous cropping due to increased soil temperature and water content that enhanced microbial activity and absence of crops to utilize mineralized nitrogen during fallow. Using partial accounting of CO_2 sources and sinks, Liebig et al. [63], however, did not found significant difference in net GWP between alternate-year fallow and continuous cropping in North Dakota. Perennial crops can reduce net GWP and GHGI compared with annual crops [7, 44, 50] due to higher root biomass production [64, 65] and increased soil carbon sequestration [55]. Because land under perennial crops is not tilled and perennial crops are not applied with fertilizers, herbicides, and pesticides, GHG emissions are usually lower with perennial than annual crops [4].

Sainju [56] found that changes in net GWP and GHGI due to crop rotation vs. monocrop and corn-soybean vs. continuous corn decreased with increased duration of experiment, but increased due to annual vs. perennial cropping systems. The relationships were further improved when soil and climatic conditions were accounted for in the multiple linear regressions of net GWP and GHGI with the duration of the experiment. He observed that soil texture had a positive effect on net GWP and GHGI for cropping intensity, but negative effect on net GWP for crop rotation vs. monocrop and perennial vs. annual crop. The trend was opposite for mean air temperature while annual precipitation had small effect. Because the magnitude of carbon sequestration rate is lower and time for carbon saturation is longer for the effect of cropping systems than for tillage systems [57, 59], reduced net GWP and GHGI for increased cropping intensity and crop rotation vs. monocrop with increased duration of experiment was due to increased carbon sequestration. Sainju [56] reported that crop rotation had a greater potential to reduce net GWP and GHGI compared with monocropping in the long run, but the potential can vary for perennial vs. annual cropping systems.

2.3. Nitrogen fertilization

Nitrogen fertilizer application rate, source, and timing and method of application can influence net GWP and GHGI. Increased nitrogen fertilization rate enhanced net GWP and GHGI due to increased N_2O emissions and CO_2 emissions associated with manufacture, transport, and application of nitrogen fertilizers, regardless of cropping systems and methods of calculations [3, 7, 21, 33, 42, 55]. In a meta-analysis of 12 experiments, Sainju [56], after accounting for all sources and sinks of CO_2 emissions, reported that net GWP decreased from 0 to ≤45 kg N ha^{-1} and net GHGI from 0 to ≤145 kg N ha^{-1} and then increased with increased nitrogen fertilization rate. Using partial accounting, net GWP decreased from 0 to 88 kg N ha^{-1} and net GHGI from 0 to ≤213 kg N ha^{-1} and then increased with increased nitrogen rate. These nitrogen rates probably corresponded to crop nitrogen demand when crops used most of the soil available nitrogen, leaving little residual nitrogen in the soil that reduced N_2O emissions and therefore net GWP and GHGI. When nitrogen rates exceeded crop nitrogen demand, net GWP and GHGI increased linearly [56], suggesting that excessive application of nitrogen fertilizers can induce net GHG emissions. Similar results have been reported by several researchers [8, 66, 67]. Therefore, nitrogen fertilizers should be applied at optimum rates to reduce net GWP and GHGI while sustaining crop yields. The optimum nitrogen rates, however, depended on net GWP measured either per unit area or per unit crop yield.

Sainju [56] observed that the relationships between net GWP, net GHGI, and nitrogen rate were further improved when the duration of the experiment and soil and climatic conditions were taken into account in the multiple linear regressions. Duration of experiment and annual precipitation had positive effects, but air temperature and soil texture had negative effects on net GWP when all sources and sinks of CO_2 emissions were accounted for. With partial accounting, only air temperature had positive effect on net GWP, but other factors had negative effects. For net GHGI, the factors having negative effects were air temperature using the complete accounting of CO_2 emissions and annual precipitation and soil texture using the partial accounting.

Alder et al. [58] reported that anhydrous ammonia reduced net GHGI compared with urea, urea ammonium nitrate, and polymer-coated urea under corn, wheat and switchgrass due to lower energy requirement for fertilizer production. They found that polymer-coated urea reduced net GHGI by slowly releasing nitrogen to the soil and reducing indirect N_2O emissions compared with urea ammonium nitrate. Little is known about the placement and methods of nitrogen fertilizer applications on net GWP and GHGI, although various results have been reported on N_2O emissions using these practices [23–29]. More research is needed about the effects of source, placement, and timing of nitrogen fertilizer application on net GWP and GHGI.

2.4. Other fertilizers

Application of combination of nitrogen, phosphorus, and potassium increased net GWP compared with no application and net GWP further increased as these nutrients were applied with a combination of inorganic fertilizer, green manure, and farmyard manure, although total amount of nutrients applied from various sources were similar under rice in China and India [43, 46]. They found that increased substrate availability from fertilizers and organic amendments increased N_2O and CH_4 emissions and therefore net GWP. Shang et al. [46], however, found lower net GHGI with these nutrient applications than without due to increased crop yield. Adviento-Borbe et al. [48] also observed increased net GWP and GHGI with combined application of nitrogen, phosphorus, and potassium compared with no application under corn in Nebraska.

2.5. Miscellaneous practices

Burning of crop residue increased net GWP and GHGI compared with residue retained in the soil due to reduced carbon input and soil carbon sequestration in upland crop production [40, 47]. Sainju et al. [55] found that irrigation increased net GWP and GHGI compared with no irrigation due to lower soil carbon sequestration as a result of increased carbon mineralization and loss of water soluble carbon from increased soil water availability. Under lowland rice, Li et al. [68] found that midseason and shallow drainage reduced net GWP and GHGI by 21–205% compared with continuous flooding. Under upland rice where flooding is minimized, they found that drainage reduced net GWP and GHGI from 17 to 322% compared with no drainage. They also found that application of nitrogen fertilizer and straw in flooded rice reduced net GWP and GHGI from 16 to 91% compared with no application, but net GWP increased by 18% by using slow N release fertilizer compared with normal nitrogen fertilizer.

2.6. Combined management practices

Using combined effects of tillage, crop rotation, and nitrogen fertilization rates, various researchers [3, 7, 49] found that net GWP and GHGI were lower with no-tillage continuous corn with reduced nitrogen rate than conventional tillage corn-soybean rotation with recommended nitrogen rate. They attributed this to increased soil carbon sequestration and reduced N_2O emissions, as corn used most of nitrogen during growth, leaving little soil residual nitrogen. They found that soybean increased N_2O emissions compared with corn, thereby increasing net GWP and GHGI with corn-soybean compared with corn. Similarly, Adviento-Borbe et al. [48] reported that net GWP and GHGI were lower with lower rates of nitrogen, phosphorus, and potassium applied to continuous corn than lower or higher rates applied to corn-soybean. Johnson et al. [51] reported that minimum till diversified crop rotation with appropriate rates of nitrogen, phosphorus, and potassium reduced net GWP and GHGI compared with conventional tillage with less diversified crop rotation and high rates of nutrients. In small grain cropping systems, Sainju et al. [9, 55] observed that net GWP and GHGI were lower with no-tillage malt-barley pea with reduced nitrogen fertilization rate than conventional tillage continuous malt barley or malt barley-fallow with recommended nitrogen rate. They attributed this to increased soil carbon sequestration, reduced N_2O emissions, and sustained crop yields.

Using a meta-analysis of nine experiments, Sainju [56] reported that the improved combined management practice that included no-tillage, diversified cropping system (crop rotation, increased cropping intensity, cover crop, and perennial cropping system) and reduced nitrogen rate reduced net GWP and GHGI by 70–88% compared with the traditional combined practice that included conventional till, less diversified cropping system (monocropping, crop-fallow, no cover crop, and annual cropping system) and recommended nitrogen rate. He also found that combined management practice further reduced net GWP and GHGI compared with individual management practices. He found that changes in net GWP and GHGI due to improved vs. traditional combined management practice increased with the duration of the experiment. The relationships were further improved by including soil and climatic factors in the multiple linear regressions. Some of the possible reasons for increased net GWP and GHGI for improved vs. traditional combined management with increased duration of the experiment are: (1) high spatial and temporal variations of GHG emissions due to differences in soil and climatic conditions and management practices, (2) reduced potential for soil C sequestration with increasing duration of the experiment, (3) use of full or partial accounting of sources and sinks of GHG emissions, and (4) uncertainty in the methods of measuring GHG emissions, such as variations in type and size of static chambers, placement of chamber in the plot (row vs. inter-row or including vs. excluding plants in the chamber), time of GHG measurement during the day, and calculation of GHG fluxes (linear or nonlinear emissions with time).

When crop residue was burned compared with residue retained in the soil under wheat applied with or without nitrogen fertilizer with various tillage practices, Wang et al. [47] found that net GWP and GHGI were lower in conventional tillage wheat without nitrogen fertilizer where residue was burned than conventional tillage or no-tillage wheat with nitrogen fertilizer where reside was either burned or retained in the soil. They found that the larger impact of N_2O emissions than soil carbon sequestration on global warming potential increased net GWP and GHGI with N fertilization than without.

Using an alternative method where soil respiration and previous year's crop residue returned to the soil are used in place of soil carbon sequestration rate to calculate net GWP and GHGI, Mosier et al. [3] observed that no-tillage continuous corn with reduced nitrogen fertilization rate reduced net GWP and GHGI compared with conventional tillage corn-soybean with recommended N rate, a case similar to that calculated by the regular method above. They attributed this to increased amount of crop residue returned to the soil and grain yield. Similarly, using this method, Sainju et al. [55] found lower net GHGI in nonirrigated no-tillage barley-pea with nitrogen fertilizer than conventional tillage continuous barley with nitrogen fertilizer, a case similar to that obtained for the regular method. They, however, observed different trends for net GWP. Similarly, using the alternative method, Johnson et al. [51] found lower net GWP and GHGI in conventional tillage corn-soybean with nitrogen and phosphorus fertilizers than no-tillage continuous corn with the same fertilizers, a case different to that obtained by using the regular method. Popp et al. [54] using the alternative method, found that net GWP was lower with nonirrigated corn than irrigated and nonirrigated cotton, soybean, sorghum, irrigated rice, and nonirrigated wheat. The magnitude of net GWP and GHGI obtained by two methods can be different, but both methods showed that no-till with continuous cropping produced lower net GWP and GHGI compared with conventional tillage with crop fallow [3, 9].

2.7. Implications of management practices

These studies showed that no-tillage systems, in general, can reduce net GWP and GHGI compared with conventional tillage systems. Perennial crops can reduce net GWP and GHGI compared with annual crops and wheat can reduce net GWP and GHGI compared with rice and corn. Inclusion of legumes in rotation with nonlegumes has variable effects on net GWP and GHGI compared with continuous nonlegumes. Inclusion of fallow in the crop rotation, however, can increase net GWP and GHGI compared with continuous cropping. Crops adequately fertilized with nitrogen, phosphorus, and potassium fertilizers can reduce net GWP and GHGI compared with no fertilized treatments, but excessive nitrogen fertilization beyond crop nitrogen demand can increase net GWP and GHGI. Burning of crop residue slightly can increase net GWP and GHGI compared with residue retained in the soil, but irrigation has minor effect compared with non-irrigation. Improving drainage or using shallow flooding in rice can lower net GWP and GHGI compared with continuous flooding. Values of net GWP and GHGI measured by the regular and alternative methods are variable, depending on soil and climatic conditions and management practices. Both methods, however, showed that the improved management practice can reduce net GHG emissions compared with the traditional management practice. Changes in net GWP and GHGI due to improved vs. traditional management varied with duration of the experiment and inclusion of soil and climatic factors improved their relationships. Also, combined management practice can lower net GWP and GHGI compared with the individual practice. Net GWP and GHGI values can be more reliable by accounting full than partial sources and sinks of CO_2 emissions. Because of the limited data, further studies are needed to evaluate the effects of management practices on net GWP and GHGI.

Author details

Upendra M. Sainju

Address all correspondence to: upendra.sainju@ars.usda.gov

USDA-ARS, Northern Plain Agricultural Research Laboratory, Sidney, Montana, USA

References

[1] Cole CV, Duxbury J, Freney J, Heinemeyer O, Minami K, Mosier A, Paustian K, Rosenberg N, Sampson N, Sauerbeck D, Zhao Q. Global estimates of potential mitigation of greenhouse gas emissions by agriculture. Nutrient Cycling in Agroecosystems. 1997;**49**:221-228

[2] Intergovernment Panel on Climate Change (IPCC). Climate Change 2014: Synthesis Report. Contribution of Working Groups I, II and III to the Fifth Assessment Report of the Intergovernmental Panel on Climate Change. Geneva, Switzerland: IPCC; 2014

[3] Mosier AR, Halvorson AD, Reule CA, Liu XJ. Net global warming potential and greenhouse gas intensity in irrigated cropping systems in northeastern Colorado. Journal of Environmental Quality. 2006;**35**:1584-1598

[4] Robertson GP, Paul E, Harwood R. Greenhouse gases in intensive agriculture: Contribution of individual gases to the radiative forcing of the atmosphere. Science. 2000;**289**: 1922-1925

[5] Schlesinger WH. Carbon sequestration in soils. Science. 1999;**284**:2095-2097

[6] West TO, Marland G. A synthesis of carbon sequestration, carbon emissions, and net carbon flux in agriculture: Comparing tillage practices in the United States. Agriculture, Ecosystems and Environment. 2002;**91**:217-232

[7] Mosier AR, Halvorson AD, Peterson GA, Robertson GP, Sherrod L. Measurement of net global warming potential in three agroecosystems. Nutrient Cycling in Agroecosystems. 2005;**72**:67-76

[8] Li B, Fan CH, Zhang H, Chen ZZ, Sun LY, Xiong ZQ. Combined effects of nitrogen fertilization and biochar on the net global warming potential, greenhouse gas intensity, and net ecosystem budget in intensive vegetable agriculture in Southeastern China. Agriculture, Ecosystems and Environment. 2015;**100**:10-19

[9] Sainju UM, Stevens WB, Caesar-TonThat T, Liebig MA, Wang J. Net global warming potential and greenhouse gas intensity influenced by irrigation, tillage, crop rotation, and nitrogen fertilization. Journal of Environmental Quality. 2014;**43**:777-788

[10] Sainju UM, Wang J, Barsotti JL. Net global warming potential and greenhouse gas intensity affected by cropping sequence and nitrogen fertilization. Soil Science Society of America Journal. 2014;**78**:248-261

[11] Lemke RL, Izaurralde RC, Nyborg M, Solberg ED. Tillage and nitrogen source influence soil-emitted nitrous oxide in the Alberta parkland region. Canadian Journal of Soil Science. 1999;**79**:15-24

[12] Ogle SM, Swan A, Paustian K. No-till management impacts on crop productivity, carbon input, and soil carbon sequestration. Agriculture, Ecosystems and Environment. 2012;**149**:37-49

[13] Angers DA, Ericksen-Hamel NS. Full-inversion tillage and organic carbon distribution in a soil profile: A meta-analysis. Soil Science Society of America Journal. 2008;**72**: 1370-1374

[14] Luo Z, Wang E, Sun OJ. Can no-till stimulate carbon sequestration in agricultural soils? A meta-analysis of paired experiments. Agriculture, Ecosystems and Environment. 2010;**139**:224-231

[15] Sainju UM, Jabro JD, Caesar-TonThat T. Tillage, cropping sequence, and nitrogen fertilization effects on dryland soil carbon dioxide emission and carbon content. Journal of Environmental Quality. 2010;**37**:98-106

[16] Sainju UM, Allen BL, Caesar-TonThat T, Lenssen AW. Dryland soil carbon and nitrogen after thirty years of tillage and cropping sequence. Agronomy Journal. 2015;**107**: 1822-1830

[17] Kuo S, Sainju UM, Jellum EJ. Winter cover crop effects on soil organic carbon and carbohydrate. Soil Science Society of America Journal. 1997;**61**:145-152

[18] Halvorson AD, Wienhold BJ, Black AL. Tillage, nitrogen, and cropping system effects on soil carbon sequestration. Soil Science Society of America Journal. 2002;**66**:906-912

[19] Gregorich EG, Ellert BH, Drury CF, Liang BC. Fertilization effects on soil organic matter turnover and corn residue carbon storage. Soil Science Society of America Journal. 1996;**60**:472-476

[20] Russell AE, Laird DA, Parkin TB, Mallarino AP. Impact of nitrogen fertilization and cropping system on carbon sequestration in Midwestern mollisols. Soil Science Society of America Journal. 2005;**69**:413-422

[21] Adviento-Borbe MA, Pittelkow CM, Anders M, van Kissel C, Hill JE, Clung AM, Six J, Linquist BA. Optimal fertilizer nitrogen rates and yield-scaled global warming potential in drill-seeded rice. Journal of Environmental Quality. 2013;**42**:1623-1634

[22] Bronson KF, Mosier AR. Suppression of methane oxidation in aerobic soil by nitrogen fertilizers, nitrification inhibitors, and urease inhibitors. Biology and Fertility of Soils. 1994;**17**:263-268

[23] Burton DL, Zebarth BJ, Gillarn KM, MacLeod JA. Effect of split application of fertilizer nitrogen on N_2O emissions from potatoes. Canadian Journal of Soil Science. 2008; **88**:229-239

[24] Phillips RL, Tanaka DL, Archer DW, Hanson JD. Fertilizer application timing influences greenhouse gas fluxes over a growing season. Journal of Environmental Quality. 2009;**38**: 1569-1579

[25] Zebarth BJ, Rochette P, Burton DL, Price M. Effect of fertilizer nitrogen management on N_2O emissions in commercial corn fields. Canadian Journal of Soil Science. 2008;**88**: 189-195

[26] Drury CF, Reynolds WD, Tan CS, Welacky TW, Calder W, McLaughlin NB. Emissions of nitrous oxide and carbon dioxide: Influence of tillage type and nitrogen placement depth. Soil Science Society of America Journal. 2006;**70**:570-581

[27] Fujinuma R, Venterea RT, Rosen C. Broadcast urea reduces N_2O but increases NO emissions compared with conventional and shallow applied anhydrous ammonia in a coarse-textured soil. Journal of Environmental Quality. 2011;**40**:1806-1815

[28] Hosen Y, Paisancharoen K, Tsuruta H. Effects of deep application of urea on NO and N_2O emissions from an Andisol. Nutrient Cycling in Agroecosystems. 2002;**63**:197-206

[29] Liu X, Mosier A, Halvorson A, Zhang F. The impact of nitrogen placement and tillage on NO, N_2O, CH_4; and CO_2; fluxes from a clay loam soil. Plant and Soil. 2006;**280**: 177-188

[30] Thornton FC, Bock BR, Tyler DD. Soil emissions of nitric oxide and nitrous oxide from injected anhydrous ammonium and urea. Journal of Environmental Quality. 1996;**25**: 1378-1384

[31] Venterea RT, Burger M, Spokas KA. Nitrogen oxide and methane emissions under varying tillage and fertilizer management. Journal of Environmental Quality. 2005;**34**:1467-1477

[32] Akiyama H, Yan X, Yagi K. Evaluation of effectiveness of enhanced-efficiency fertilizers as mitigation options for N_2O and NO emissions from agricultural soils: Meta-analysis. Global Change Biology. 2010;**16**:1837-1846

[33] Halvorson AD, Del Grosso SJ, Alluvione F. Nitrogen source effects on nitrous oxide emissions from irrigated no-till corn. Journal of Environmental Quality. 2010;**39**:1554-1562

[34] Hyatt CR, Venterea RT, Rosen CJ, McNearney M, et al. Polymer-coated urea maintains potato yields and reduces nitrous oxide emissions in a Minnesota loamy sand. Soil Science Society of America Journal. 2010;**74**:419-428

[35] Robertson GP. Keeping track of carbon. Science. 1999;**285**:1849

[36] Decock C. Mitigating nitrous oxide emissions from corn cropping systems in the Midwestern U.S.: Potential and data gaps. Environmental Science & Technology. 2014;**48**: 4247-4256

[37] Sainju UM, Stevens WB, Caesar-TonThat T, Liebig MA. Soil greenhouse gas emissions affected by irrigation, tillage, crop rotation, and nitrogen fertilization. Journal of Environmental Quality. 2012;**41**:1774-1786

[38] Basche AD, Miguez FE, Kaspar TC, Castellano MJ. Do cover crops increase or decrease nitrous oxide emissions? A meta-analysis. Journal of Soil and Water Conservation. 2014; **69**:471-482

[39] Curtin D, Wang H, Selles F, McConkey BG, Campbell CA. Tillage effects on carbon dioxide fluxes in continuous wheat and fallow-wheat rotations. Soil Science Society of America Journal. 2000;**64**:2080-2086

[40] Robertson GP, Grace PR. Greenhouse gas fluxes in tropical and temperate agriculture: The need for a full-cost accounting of global warming potentials. Environment, Development and Sustainability. 2004;**6**:51-63

[41] Linquist B, van Groenigen KJ, Adviento-Borbe MA, Pittelkow C, van Kissel C. An agronomic assessment of greenhouse gas emissions from major cereal crops. Global Change Biology. 2012;**18**:194-209

[42] Pittelkow CM, Adviento-Borbe MA, Hill JE, Six J, van Kissel C, Linquist BA. Yield-scaled global warming potential of annual nitrous oxide and methane emissions from continuously flooded rice in response to nitrogen input. Agriculture, Ecosystems and Environment. 2013;**177**:10-20

[43] Bhatia A, Pathak H, Jain N, Singh PK, Singh AK. Global warming potential of manure-amended soils under rice-wheat system in the indo-Gangetic Plains. Atmospheric Environment. 2005;**39**:6976-6984

[44] Ruan L, Robertson GP. Initial nitrous oxide, carbon dioxide, and methane costs of converting conservation reserve program grassland to row crops under no-till vs. conventional tillage. Global Change Biology. 2013;**19**:2478-2489

[45] Piva JT, Diekow J, Bayer C, Zanatta JA, de Moraes A, Pauletti V, Tomazi M, Pergher M. No-till reduces global warming potential in a subtropical Ferrasol. Plant and Soil. 2012; **361**:359-373

[46] Shang Q, Yang X, Gao C, Wu P, Liu J, Xu Y, Shen Q, Zou J, Gao S. Net annual global warming potential and greenhouse gas intensity in Chinese double rice cropping systems: A 3 year field measurement in the long-term fertilizer experiments. Global Change Biology. 2011;**17**:2196-2210

[47] Wang W, Dalal RC, Reeves SH, Butterbach-Bahl K, Kiese R. Greenhouse gas fluxes from an Australian subtropical cropland under long-term contrasting management regimes. Global Change Biology. 2011;**17**:3089-3101

[48] Adviento-Borbe MA, Haddix ML, Binder DL, Walters DT, Dobermann A. Soil greenhouse gas fluxes and global warming potential in four high-yielding maize systems. Global Change Biology. 2007;**13**:1972-1988

[49] Archer DW, Halvorson AD. Greenhouse gas mitigation economics for irrigated cropping systems in northeastern Colorado. Soil Science Society of America Journal. 2010;**74**: 446-452

[50] Barsotti JL, Sainju UM, Lenssen AW, Montagne C, Hatfield PG. Net greenhouse gas emissions affected by sheep grazing in dryland cropping systems. Soil Science Society of America Journal. 2013;**77**:1012-1025

[51] Johnson JMF, Archer DW, Weyers SL, Barbour NW. Do mitigation strategies reduce global warming potential in the northern U.S. Cornbelt? Journal of Environmental Quality. 2011;**40**:1551-1559

[52] Kim S, Dale BE. Cumulative energy and global warming impact from the production of biomass for biofuel products. Journal of Industrial Ecology. 2004;**7**:147-162

[53] Kim S, Dale BF, Jenkins R. Life-cycle assessment of corn grain and corn stover in the United States. International Journal of Life Cycle Assessment. 2009;**14**:160-174

[54] Popp M, Nalley L, Fortin C, Smith A, Brye K. Estimating net carbon emissions and agricultural response to potential carbon offset policies. Agronomy Journal. 2011;**103**:1132-1143

[55] Sainju UM, Stevens WB, Caesar-TonThat T. Soil carbon and crop yields affected by irrigation, tillage, crop rotation, and nitrogen fertilization. Soil Science Society of America Journal. 2014;**78**:936-948

[56] Sainju UM. A global meta-analysis on the impact of management practices on net global warming potential and greenhouse gas intensity from cropland soils. PLoS One. 2016;**11**:1, e0148527-26. DOI: 10.1371/journal.pone.0148527

[57] Six J, Ogle SM, Breidt FJ, Conant RT, Mosier AR, Paustian K. The potential to mitigate global warming with no-tillage management is only realized when practiced in the long term. Global Change Biology. 2004;**10**:155-160

[58] Adler PK, Del Grosso SJ, Inman D, Jenkins RE, Spatari S, Jhang Y. Mitigation opportunities for life-cycle greenhouse gas emissions furing feedstock productions across heterogeneous landscapes. In: Liebig A, Franzluebbers AJ, Follett RF, editors. Managing Agricultural Greenhouse Gases. Academic Press; 2012. pp. 203-219

[59] West TO, Post WM. Soil organic carbon sequestration rates by tillage and crop rotation: A global data analysis. Soil Science Society of America Journal. 2002;**66**:1930-1946

[60] Jensen ES, Peoples MB, Boddey RM, Gresshoff PM, Hauggard-Nelson H, Alves BJR, Morrison MJ. Legumes for mitigation of climate change and the provision of feedstock for biofuels and biorefineries. Agronomy for Sustainable Development. 2012;**32**:329-364

[61] Lemke RL, Zhong Z, Campbell CA, Zentner R. Can pulse crops play a role in mitigating greenhouse gases from north American agriculture? Agronomy Journal. 2007;**99**: 1719-1725

[62] Peterson GA, Halvorson AD, Havlin JL, Jones OR, Lyon DG, Tanaka DL. Reduced tillage and increasing cropping intensity in the Great Plains conserve soil carbon. Soil and Tillage Research. 1998;**47**:207-218

[63] Liebig MA, Tanaka DL, Gross JR. Fallow effects on soil carbon and greenhouse gas flux in central North Dakota. Soil Science Society of America Journal. 2010;**74**:358-365

[64] Ma Z, Wood C, Bransby DI. Impacts of soil management on root characteristics of switchgrass. Biomass and Bioenergy. 2000;**18**:105-112

[65] Qin R, Stamp P, Richner W. Impact of tillage on root systems of winter wheat. Soil Science Society of America Journal. 2004;**96**:1523-1530

[66] Huang T, Gao B, Christie P, Ju X. Net global warming potential and greenhouse gas intensity in a double-cropping cereal rotation as affected by nitrogen and straw management. Biogeosciences. 2013;**10**:7897-7911

[67] Ma YC, Kong XW, Yang B, Yang XL, Yang JC, Xiong ZQ. Net global warming potential and greenhouse gas intensity of annual rice-wheat rotations with integrated soil-crop system management. Agriculture, Ecosystems and Environment. 2013;**164**:200-219

[68] Li C, Salas W, Angelo B, Rose S. Assessing alternatives for mitigating net greenhouse gas emissions and increasing yields from rice production in China over the next twenty years. Journal of Environmental Quality. 2006;**35**:1554-1565

Chapter 6

Low Carbon Technologies for Agriculture in Dryland: Brazilian Experience

Vanderlise Giongo, Alessandra Monteiro Salviano, Francislene Angelotti, Tatiana Taura, Luiz Fernando Carvalho Leite and Tony Jarbas Ferreira Cunha

Additional information is available at the end of the chapter

http://dx.doi.org/10.5772/intechopen.72363

Abstract

Anthropogenic activities have altered the atmospheric composition since the industrial era, especially with the increasing greenhouse gas emission due to fossil fuel combustion, cement production, and land-use change. The Brazilian semiarid, covering approximately 969.589 km^2 with 21 million people, region has 1.6 million agricultural establishments and 95% are classified as family farms. The typical agricultural systems are characterized by high grazing density, slash and burn practices, and fruits and legumes by irrigated monocultures. Consequently, soil degradation occurs due unsustainable soil management, decreasing soil carbon stock, and the biodiversity. The soil carbon depletion is also associated with saline, water, and thermal stresses. Saline, water, and thermal stresses in dryland, the impact of the land-use change associated with climate change, and few technological resources available for use in agricultural systems are the main reasons responsible for low productivity in the Brazilian semiarid region. Low-cost agricultural practices can contribute to build healthy and sustainable agroecosystems: among these, the selection of plant species tolerant to saline, water, and thermal stresses, the use of rhizobial inoculants, adoption of no-tillage, sowing green manure, and adoption of technologies to stock water to improve its efficiency and productivity.

Keywords: climate change, land-use change, technologies, agroecosystems design

1. Introduction

All the governments of the world have been concerned about climate change and how they can ensure access to sufficient food, water, and energy resources to safeguard human well-being

[1–3]. The anthropogenic activities have altered the atmospheric composition since the industrial era, especially by the increasing of greenhouse gas emission due to fossil fuel combustion, cement production, land-use change, and land use [2, 4, 5]. Land use and land-use changes affect soil carbon stocks, which are particularly important because they are the largest and most stable active compartment of the planet that can be handled [5–7]. In Brazil, land use and land-use changes, caused by deforestation or agricultural practices, can have a large participation on total national greenhouse gas emission (GHG) having reached 58% of all CO_2eq emitted in 2005 [7].

The most optimistic greenhouse gas emission (GHG) scenarios projected that planet temperature will increase at least by 2°C until the year 2100 [8, 9]. Managing with different climate change scenarios, in December 2015, Parties of the United Nations Framework Convention on Climate Change (UNFCCC) adopted the Paris Agreement to manage climate change. The Paris Agreement stipulated that it is necessary to limit temperature increase to 1.5°C above preindustrial levels by 2050 [10, 11]. Although, even if all national commitments in the Paris Agreement are accomplished, mean planet temperature is likely to increase at least by 2.6–3.1°C until the year 2100 compared to preindustrial levels [12]. Even at the face of difficulties, many Parties formulated and submitted Intended Nationally Determined Contributions or INDCs that outline the post-2020 climate action plans they intend to take under the Paris Agreement. On that basis, Brazil undertook to reduce until 2025 the greenhouse gas emissions by 37% below 2005 levels and until 2030 reduce greenhouse gas emissions to 43% below 2005 levels. All regions of the country will be developing local actions to achieve the national goal. As much of the country's emissions are linked to land-use change and agriculture, these themes are prominent in research, extension, and public policy actions [13, 14].

In that way, one of the main programs established by the country that will support the reach of the INDCs is the Sectorial Plan of Mitigation and Adaptation to the Climate Change for the Consolidation of a Low Carbon Economy in agriculture—ABC Plan—created in 2010 and whose objectives are to promote the reduction of greenhouse gas emissions in agricultural activities; reduce deforestation; increase agricultural production on a sustainable basis; adapt rural properties to environmental legislation; expand the area of cultivated forests; and stimulate the recovery of degraded areas. Thus, the ABC Plan represents a set of applied technologies in agriculture and livestock, able to promote the reduction of GHG emissions by improving management practices and increasing carbon retention in soil and vegetation, while raising rural farming income [15, 16].

The Brazilian semiarid region, covering approximately 969,589 km^2, is situated in the Northeastern part (**Figure 1**), with unique native dryland vegetation and adapted to the periodic droughts called Caatinga. With 21 million people, this region has 1.6 million agricultural establishments and 95% are classified as family farms [17, 18]. Much of this population still seeks their livelihood in agropastoral activities and based on natural resources existing on or around their properties. Consequently, the land-use change from woody plants used for energy production, together with the conversion of use aimed at agricultural production, is responsible for the removal of 46.38% of the Caatinga vegetation [10, 19]. The native vegetation, to produce firewood and charcoal, is more than 30% of the energy matrix of the Brazilian semiarid region, and the demand for these products increases the deforested areas to improve farmers' income.

The climatic scenarios point out an increase in average air temperature up to 4.8°C and a 50% reduction in rainfall distribution by the end of the century (2071–2100) [8]. Rising air

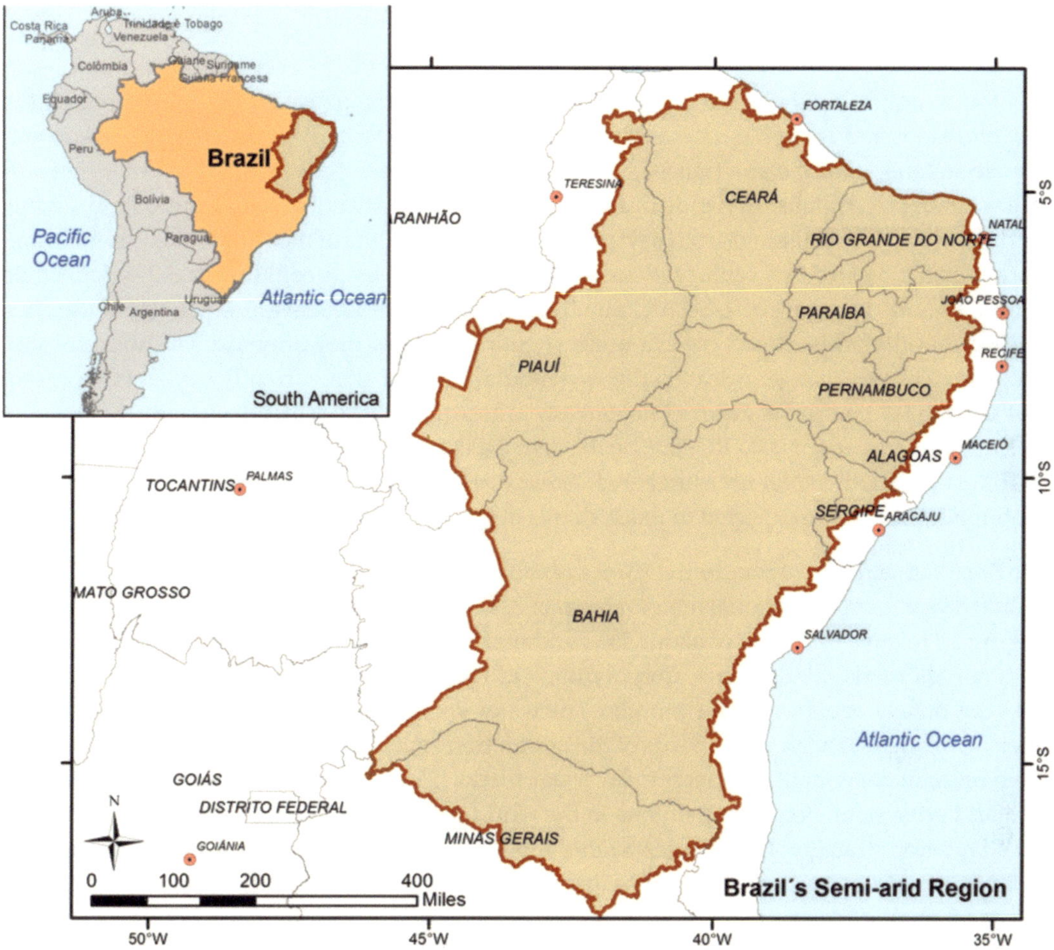

Figure 1. Map of Brazilian semiarid region.

temperatures can intensify hydric deficiency, affecting the availability of water for human consumption and for the rain-dependent agricultural activities. The increase of drought in the Northeast and changes in the characteristics and distribution of vegetation warn the risk of aridization of the region [20]. With these characteristics, the semiarid region is the most vulnerable Brazilian region to climate changes due to the increased difficulty of accessing water, food, and energy and due to the economic and social crisis [21].

This chapter aims to demonstrate the importance of low carbon agriculture to ensure access to food, energy, and water and to ensure human well-being facing the global climate change scenarios in the Brazilian semiarid region. In this regard, we understand that the way forward is to provide tools (technologies, products, processes, and knowledge) so that government, farmers, and private initiatives can create structures and links to implement sustainable agroecosystems, integrating concepts of resilience, adaptation, mitigation, and transformation of the biosphere in a unique approach to reflective and collaborative science.

2. Dryland characteristics and fragilities

It is important to highlight the large asymmetry between the geographic distributions of the CO_2 emissions and warming rates and work with indicators of corn yield, floods, drought, and climate suitability for malaria transmission. Researches point out over the past century that surface warming over global drylands has been 20–40% higher than that over humid lands, while anthropogenic CO_2 emissions from drylands have been only 30% of those generated from humid lands. For the twenty-first century, warming over global drylands could reach 3.2–4.0°C, while in humid lands warming of 2.4–2.6°C due to anthropogenic CO_2 emissions [22]. With this, it is evident that drylands should receive more attention because they are most sensitive and vulnerable to climate change. Additionally, we should consider that currently drylands represent 45.4% of the 147,000 km^2 of Earth's total terrestrial area [23]. Studies, based on aridity data analysis simulated under scenario RCP8.5, by the year 2100, estimate that the drylands will represent 56.10% of the Earth's total terrestrial area. Areas covered by hyperarid, arid, semiarid, and dry subhumid systems are expected to reach shares of 12.6, 14.9, 20.3, and 8.3%, respectively [22].

The Brazilian semiarid region, with 11% of the national territory (**Figure 1**), has average insolation of 2800 h $year^{-1}$, average annual temperatures of 23–27°C, average evaporation of 2000 mm $year^{-1}$ (**Figure 2**), relative humidity of about 50%, and maximum annual rainfall of 800 mm (**Figure 2**), with rainfall marked by scarcity, irregularity, and concentration of rainfall in a short period of the year, on average, from 3 to 4 months. The water volumes stored in lakes and reservoirs are often insufficient to achieve the needs of the population and to feed the animals [24], and farmers often become dependent on government water trucks [25, 26]. The Caatinga Biome is the main semiarid ecosystem. Botanically unique in the world, its vegetation is resilient due to the different adaptation strategies to the water, saline, and thermal stresses, due to the high temperatures and periodic droughts that characterize this region. Expressed in a multifaceted mosaic of fragments of small dimensions, with soils of extremely different characteristics occurring in close proximity, the soil diversity of the semiarid region is the largest in Brazil [27].

The land use and occupation in the semiarid region happened due to the expansion of the area for cattle breeding in the eighteenth century, the period of colonial Brazil. However, this occupation occurred in a disorderly way and without taking into consideration the fragility of natural resources. The management of these animals was carried out in an ultra-extensive way (animals released in the open field and having as food source the tree, shrub, and herbaceous species of the Caatinga) [28]. Thus, cattle ranching was responsible for the demographic occupation of the semiarid region, giving rise to settlements, which later became large cities. At that time, subsistence agriculture also began, characterized by exploitation during the rainy season, in small orchards surrounded by sticks, with cassava, corn, and beans.

The livestock and subsistence crops are still the main land uses. The most diverse arrangements can be observed in the Brazilian semiarid region, but all of them stand out due to the small area organized; approximately 95% of the establishments are family farms (**Table 1**) [17, 28]. The existing demand for firewood and charcoal extends the deforested areas of native vegetation to improve income from the sale of wood. The production of firewood and charcoal from native vegetation constitutes more than 30% of the energy matrix [29]. Subsistence agriculture, with cassava, beans, maize, and several annual crops, occupies small spaces and does not promote a

http://dx.doi.org/10.5772/intechopen.72363

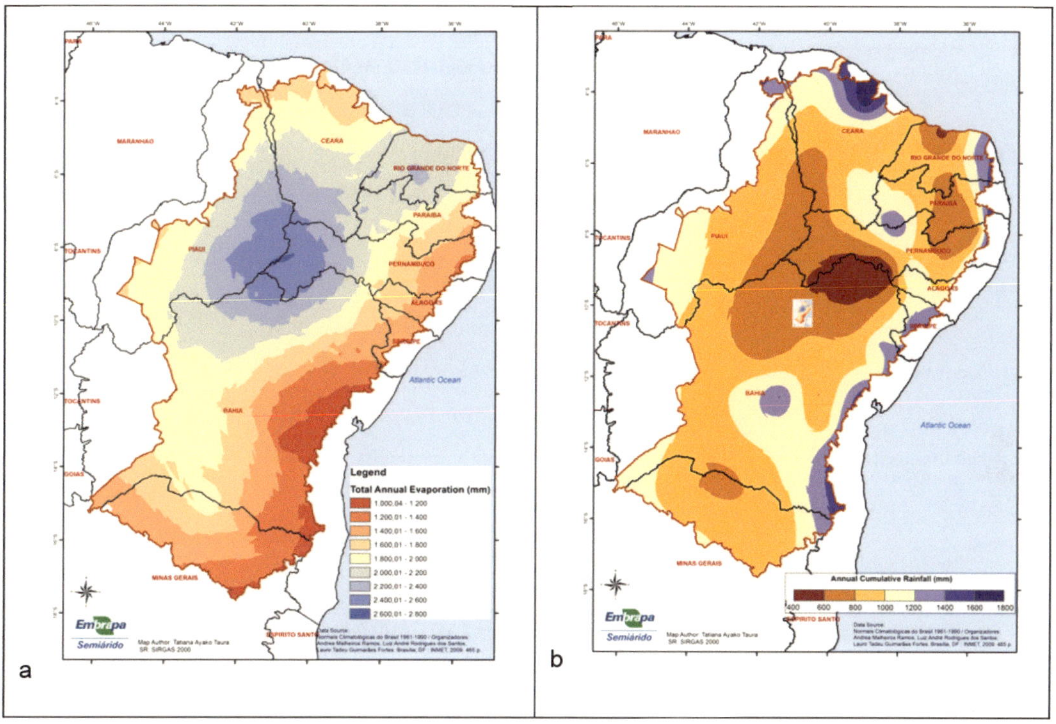

Figure 2. Evaporation (a) and precipitation (b) maps of the Brazilian semiarid region.

deforestation front. Small farming or pasture plots are exploited for a few years and abandoned for longer periods of time [30]. However, its itinerant characteristic increases deforested areas in a pulverized form. To a lesser extent, irrigation projects that produce fruits for export are responsible for the deforestation of the Caatinga [27, 28].

The livestock, characterized by overgrazing, itinerant subsistence agriculture, wood extraction, poorly designed irrigation projects, and increase of severe drought are contributing to enhance the water, thermal, and salt stresses (**Figure 3**).

Climate change, land-use change, and land use are expanding degraded and desertified areas, leading to loss of biodiversity, soil carbon stock, and vegetation [6], and, in general, degrading physical and chemical properties of soil that does not support primary productivity [6, 24, 31–33]. However, Brazilian semiarid region presents a great diversity of soil types and land use expressed in a multifaceted mosaic of small fragments occurring in the vicinity [27] (**Figure 4**).

Studies on the impact of land-use change and climate change on the Brazilian semiarid region show that resilience can be drastically affected by atrophic action and climate change. In a study by means of time series of difference vegetation index satellite normalized derivative (NDVI) 2008–2013 and weather data, the cleared areas had significantly lower normalized difference vegetation index (NDVI) and greening delay in response to precipitation. On the other hand, strictly protected areas presented higher productivity and considerable resilience at low levels of precipitation compared to sustainable use or unprotected areas [34]. Changes in the characteristics and distribution of vegetation associated with increased drought and

Production systems	Characteristics
Survival farming	- Crops for high consumption (rice, corn, beans, and fava beans)
	- Have no animal breeding
Subsistence farming	- Survival cultures
	- Maximum of 3 ha in crops of commercial value
Commercial agriculture	> 3 ha of commercial agriculture
Livestock production	- Maximum of five animal units
	- Self-consumption crops
Diversified livestock subsistence	- Up to five animal units
	- Maximum of 3 ha of commercial crops
Diversified livestock with commercial agriculture	- Up to five animal units
	> 3 ha of commercial crops
Livestock	- Crops for self-consumption
	- Five animal units
	- Produce <7000 L milk/year
Diversified livestock	- Up to five animal units
	- Maximum of 3 ha of commercial crops
	- Produce <7000 L milk/year
Livestock with commercial agriculture	- >5 units animal
	- Maximum 7000 L milk/year
	- More than 3 ha of commercial crops
Livestock milk	- >5 animal units
	- Crops for self-consumption
	- Produce >7000 L/year milk
Diversified livestock milk	- >5 animal units
	- 3 ha of commercial crops
	- Produce >7000 L/year milk
Livestock milk with commercial agriculture	- >5 animal units
	- >3 ha of commercial crops
	- Produce >7000 L/year milk

Adapted from [28].

Table 1. Types of production systems in the Brazilian semiarid region.

temperatures and atrophic action point to the risk of degradation and aridization of the region [20]. These studies allow us to verify two important questions: first, the importance of preserved areas and second, the need to develop agricultural systems and to use natural

Figure 3. Images of the Caatinga in the dry period (a), current model of goat breeding, exploring native vegetation and degraded pastures (b), area in which vegetation was removed and burnt (c), and area cultivated with forage palm (d). Source: Embrapa image bank.

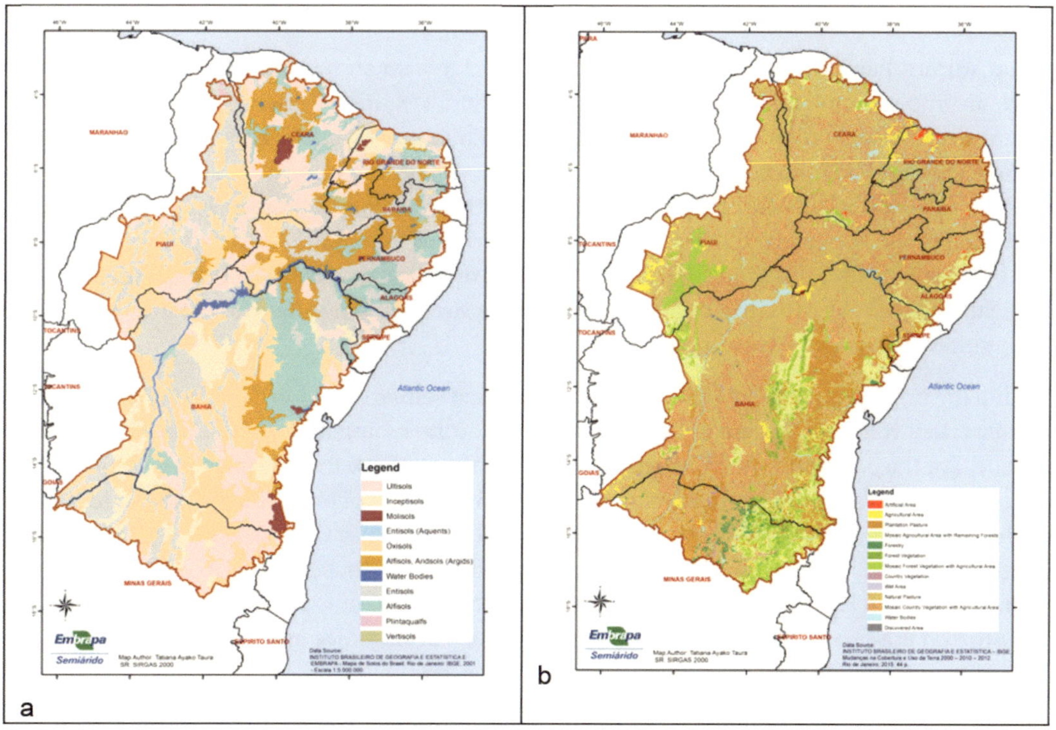

Figure 4. Soil (a) and vegetation (b) maps of the Brazilian semiarid region.

resources that do not affect biodiversity and the regeneration capacity of Caatinga and that do not put pressure on the preserved areas while adapting to the impacts of climate change.

Impacts due to temperature rise and precipitation anomalies can cause socioeconomic and ecological damage. From the economic point of view, climatic variations have a direct impact on agricultural production, representing a challenge for food security. Similarly, in semiarid regions, climate change can further aggravate and accelerate the process of degradation/desertification, affecting ecosystem function and biodiversity, causing loss of landscape heterogeneity [35, 36] in addition to increasing emissions of greenhouse gases and decreasing the capacity to store carbon in soil and vegetation.

3. Direct impacts of climate change on productive systems

The increase in temperature and changes in precipitation patterns may cause significant impacts on the different types of production systems in the semi-arid region. This is because the air temperature is one of the main climatic elements for the growth and development of plants. It is known that high temperatures can reduce the metabolic activity and increase breathing [37]. Cowpea, for example, is a crop of great socioeconomic importance for the semiarid region and this leguminous develops well between temperatures of 20 and 30°C, being that high temperatures cause spontaneous abortion of flowers and the retention of pods in the plant [38]. Thus, in places where the average temperature varies between 26 and 27°C, an increase of 4.8°C [8] could change the development of the plants, harming their production.

Some studies based on climatic zoning of crops have also shown that climate change may have an impact on semiarid agricultural production, negatively interfering with the yield of some traditional crops of family agriculture, such as cassava [39]. This is because, climate change changes the hydrographic cycle, generating changes in hydric availability. And in the cassava case, the plants can be grown in temperatures ranging between 16 and 38°C; however, due to the scenarios of the dry season increase, this cultivation may have an increase in the risk area for its production. Thus, the amount of soil water available for cassava may be a negative factor if the hydric deficit occurs during the first 5 months after planting, needing adaptation measures to avoid possible losses [39].

Within the production systems existing in the Brazilian semiarid region, livestock farming plays an important role both for income generation and for the maintenance of families in the countryside. For the goats and sheep, about 80% of the properties use the Caatinga as a source of forage. In drought years, other practices such as the consumption of grains/pods, hay, fodder palm in the trough, and buffel grass are adopted for food management. However, irregular use and insufficient offer are the major bottlenecks of the current production system [40]. With climate change scenarios, hydric deficiency can directly affect the yield of forage species, thus increasing pressure on the Caatinga and intensifying also the lack of water for animal consumption.

A study carried out in the state of Pernambuco, in the Brazilian semiarid region, verifying the impact of land-use change on the carbon stock in different ones, showed that the carbon stock is higher in soils with higher clay content and these are more sensitive to use changes in the first layer (0–0,1 m). However, in the layer of 0–60 cm, land-use change has a significantly greater impact on sandy soils when compared to clayey soils (**Table 2**) [41].

In the Brazilian semiarid region, sandy and low carbon soils predominate [30, 41] (**Table 3**). Thus, facing climate change scenarios, primary productivity would be compromised and carbon stocks would drastically reduce because of the predominant soil characteristics, aggravating both emissions and carbon storage capacity.

The impacts of climate change are not only restricted to agricultural production. The reduction in soil water availability may promote the vegetation replacement from semiarid regions by arid vegetation regions [20]. In this sense, some studies indicate biodiversity losses and increase of vulnerable areas to desertification. For the population that uses native plants as a source of animal feed and medicinal use and that even explores the diversity of fruit species for human consumption and to increase family income, the advent of climate change may be more of a challenge that needs to be understood to avoid the process of environmental degradation. This is because the climatic variability in semiarid regions coupled with human activities has led to the loss of biological and economic productivity of agricultural lands, pastures, and native forest areas, losing the ability to recover. Thus, understanding how changes in climate can influence the distribution of species, as well as establishment and regeneration, is a challenge that needs to be investigated to maintain the sustainability of this ecosystem.

The negative impact of a fall in agricultural production, with a consequent decrease in income, may apply a reduction of jobs, as well as a rural exodus. In this way, the rainwater harvesting and storage technologies implantation, the use of genetic materials resistant to drought and high temperatures, the integration of technologies, and the use of polycultures will be extremely important to achieve sustainable development of the region against the climate change, reducing risks and promoting the maintenance of family farming.

The main reasons responsible for the low productivity in the Brazilian semiarid region are water, thermal, and saline stresses in dryland, the impact of the land-use change associated with climate change, as well as few technological resources available for use in agricultural systems. The use of integrated technologies is important to increase the carbon stock and to mitigate climate change, increasing the productivity of agroecosystems in Brazilian's drylands.

Land use	**Carbon (Mg ha^{-1})**			
	Acrisol	**Ferralsols**	**Leptosol**	**Planosol**
Dense Caatinga	63.8 (5.5*)	47.2 (6.8)	54.5 (11.8)	26.2 (4.2)
Open Caatinga	45.9 (4.6)	39.7 (6.7)	39.1 (4.5)	26.2 (3.1)
Pasture	51.3 (10.4)	39.4 (4.6)	51.4 (0.7)	17.6 (2.8)
Agriculture	56.0 (5.7)	34.4 (5.6)	22.2 (1.5)	13.3 (0.7)

*Mean standard error.

Pernambuco, Brazil, 2014. Adapted from [41].

Table 2. Carbon stocks (mg ha^{-1}) in the top 0–60 cm soil layer under different land uses and soil types.

Soil type		ON*	Organic carbon	Clay	Silt	Sand	Area	
FAO-UNESCO	USDA			g kg^{-1}			km^2	%
Ferralsols	Oxisols	41	9.7	250	130	620	203.614	21
Leptosols	Lithic ...Orthents	45	10.4	132	250	618	184.222	19
—	Lithic...Psamments						—	—
Acrisols	Ultisols	90	8.9	147	157	696	145.438	15
Luvisols	Ultisols	47	11.8	176	258	566	126.047	13
Planosol	Albic suborder	68	7.4	105	188	707	38.784	4
Regosols	Orthents, Psamments	20	4.9	37	105	858	38.784	4
Cambisols	Inceptisols	13	12.2	295	212	493	38.784	4
Vertisols	Vertisols	16	12.3	374	238	388	9.696	1
Others							184.222	19
Total							969.589	100

Adapted from [27].
*ON, observation number.

Table 3. Soil types according to two different soil classification systems and relationship between carbon content and particle in surface horizons of the main types of soils in Brazilian semiarid region.

We will describe in this chapter some low-cost agricultural practices that can contribute to build healthy and sustainable agroecosystems. Among these, we will focus on the selection of plant species tolerant to the saline, water, and thermal stress, use of rhizobial inoculants to benefit economic and environmental impacts for leguminous crops, adoption of no-tillage systems, sowing of different species of green manure called plant mixture or plant cocktail, and technologies to stock water to improve its efficiency and productivity.

4. Low-cost agricultural practices that can contribute to build healthy and sustainable agroecosystems

4.1. Plant species tolerant to the saline, water, and thermal stresses in drylands

The plant species of the Caatinga present adaptations to face the low hydric availability, high temperatures, and salinity. The occurrence of abiotic stresses provokes biochemical and physiological responses in plants, to promote tolerance or increase their survival under adverse conditions. The plants adapted to drought, high temperatures, and salinity usually present some strategies such as succulence, dormancy, and leaves with serous layers or with capacity to store water and nutrients in specific structures of roots [42]. In this context, plant genetic resources, associated with biodiversity and biotechnology, are essential to explore new materials that make agriculture more competitive, secure, and sustainable The value of genetic resources is enormous, and their conservation, characterization, and use are fundamental for improvement programs that aim to identify species adapted to adverse conditions, such as the semiarid climate [42]. However, the strategy of identifying the mechanisms of adaptation to the stresses in native species and the incorporation to the cultivated species is quite complex and are currently linked with the association of genes, proteins, and others.

Genetic improvement of cultivated species is another important strategy aiming at tolerance to hydric deficit, temperature increases, and salt stress to guarantee the sustainability of agricultural production.

For these strategies, viability, the use of tools such as bioinformatics, systems biology, interaction ratio among them, the association of data deposited in databases, laboratory data, and field are essential. These studies require network projects, with complementary and interdisciplinary approaches, which need significant investments. To face this challenge, plant improvement programs are expanding the genetic base of prospecting and accelerating the search for novel phenotypic traits [42, 43].

In the semiarid region, some initiatives have contributed to the search for resistance genes to abiotic stresses. An example is the evaluation of cassava varieties resistant to dehydration in genetic improvement programs, searching for more productive varieties under conditions of water deficiency [44, 45]. The cultivars of guandu beans (Guandu Petrolina and Guandu Taipeiro) developed to adapt to the irregular regime of semiarid rains [46] are also noteworthy. The onion cultivar "Alfa São Francisco" was launched to resist to high temperature and it is a good option to family agriculture [47]. Through the transcriptome and proteome analysis, associated with physiological studies, we seek to understand the genetic mechanisms of *Tripogon spicatus* adaptation to drought. The results of this research may contribute to the generation of biotechnological alternatives for the improvement of cultivated plants, through the identification of genes associated with stress tolerance [48].

Soil salinization can occur by natural processes, named primary salinization, or by induced processes, named secondary salinization or anthropic. In the semiarid region, this process is accentuated due to the negative water balance most of the year with potential evapotranspiration of 2000 mm/year. Secondary salinity is a problem that affects the semiarid region of Brazil, especially in the irrigated perimeters. Studies investigating the evolution of salinity in an Argissol under irrigation in Petrolina—PE—observed that the indiscriminate use of salts in the fertilization and the excessive use of water contributed to a process of anthropic salinization in irrigated area. However, it is possible to consider that the salinization process, being in its initial phase, could be reversible, and among some measures suggested are the correction of the excess water applied, as well as avoiding the use of fertilizers with high saline indices, and the use of organic fertilization and/or green manure as a management practice [49]. Thus, in order to maintain a sustainable agriculture in the semiarid region, it is necessary to follow the chemical evolution of the soils in order to characterize the salinization process, the adoption of management practices for mitigation, and the selection of tolerant species strategies to increase productivity.

Several plants are able to grow under salinity conditions. The species *Atriplex nummularia* is a halophyte forage that shows high tolerance to salinity (>25 dS/m) [50]. In the semiarid region, this species is produced under irrigated conditions with desalination waste, producing a total of 55 t ha^{-1} an^{-1} of dry matter [50]. The good productive performance of this species allows the mobilization of soil salts and the production of firewood and forage material [51].

4.2. Economic and environmental impacts of the use of bacteria and mycorrhizal fungi

The use of efficient bacteria in the biological fixation of nitrogen and mycorrhizal fungi is a recent initiative used in the Brazilian semiarid regions, which aims to contribute to the reduction of the climate change impacts. In a scenario of climate change, the selection of efficient bacteria in the biological fixation of nitrogen and mycorrhizal fungi that increase the absorption of water and soil nutrients may contribute as important tools to help plant species and to reduce the impacts of adverse climatic conditions, such as high temperatures and low water availability [52–54].

In biological nitrogen fixation (BNF), the bacteria fix the atmospheric nitrogen in organic compounds that are used by plants, reducing the necessity to use nitrogenous fertilizers and improving the absorption of water and nutrients. The BNF, on the other hand, does not have specificity regarding the host plants; however, there are studies that indicate that BNF has ecological specificity. Species isolated from certain plant communities are more adapted to plants and to prevailing edaphoclimatic conditions, and they are therefore more adapted and able to colonize the root system and favor the development of the plant species that occur in these places [55].

Thus, the use of these techniques allows a greater production of the plants and an increase in the capacity to support environmental stresses, being able to be an additional tool in the integration of technologies for the family farming. Some published works indicate the potential use of these tools for leguminous crops [56, 57]. For cowpea, four strains of *Bradyrhizobium* sp. are currently authorized to produce inoculants in Brazil, the most widespread being the BR 3267 strain originating from Petrolina soils [58], demonstrating the potential of the region as a source of microorganisms. In addition, plant genotypes growing in the semiarid region show responsiveness to the inoculation of the bacteria used in commercial inoculants [59], which reinforces the necessity for constant prospecting of new rhizobia isolates.

4.3. No-tillage systems and the plant mixture

The no-tillage system is one of the main technologies encouraged by the ABC Plan. In the Brazilian agricultural soils, no-tillage system favors carbon sequestration, with increases of 5.2–8.5 MgC ha^{-1}, higher than the soil under conventional tillage [60]. However, the Brazilian semiarid region is the most difficult place for technology to be implemented for low-carbon agriculture. The difficulty of implementing the no-tillage system in the Brazilian semiarid region is due to climatic restrictions and cultural remains traditionally used to feed the herds [60, 61].

The use of green manures can be a viable low carbon emission technology for irrigated agriculture in the Brazilian semiarid region. The simultaneous cultivation of different green manure species is an alternative to take benefits promoted by different species [62–64]. In that way, studies in long-term experiments, using as a model for fruit the mango tree and for horticultural the melon (**Figure 5b–d**). In both systems, mango tree and melon, the simultaneous cultivation of 14 different species of green fertilizers, called plant mixture, was carried out, contemplating differentiated proportions of grasses, oilseeds, and legumes [61, 62]. The selected species were legumes (*Calopogonium mucunoides*), velvet bean (*Stizolobium aterrimum* L.),

Figure 5. Plant mixture preceding melon crop (a), t melon seedlings in plant mixture residues (b), no-tillage system of melon crop (c), and detail of the harvest phase with residues still on the soil (d). Source: Embrapa image bank.

gray-seeded mucuna (*Stizolobium cinereum* Piper and Tracy), crotalaria (*Crotalaria juncea*), rattlebox (*Crotalaria spectabilis*), jack beans ensiformis), lab-lab bean (*Dolichos lablab* L.); grasses: sesame (*Sesamum indicum* L.), corn (*Zea mays*), pearl millet (*Pennisetum americanum* L.), and milo (*Sorghum vulgare* Pers.); oil seed: pigeon pea (*Cajanus cajan* L.), sunflower (*Helianthus annuus*), castor oil plant (*Ricinus communis* L.) (**Figure 5a**). The spontaneous vegetation was composed by the predominant species: Benghal dayflower (*Commelina benghalensis* L.), purple bush-bean (*Macroptilium atropurpureum*), Florida beggarweed (*Desmodium tortuosum*), and goat's head (*Acanthospermum hispidum* DC).

Oilseeds, such as sunflower and castor oil, produce large amounts of biomass and cycling nutrients, especially nitrogen. Grasses generally contribute with relatively high amounts of phytomass, characterized by high C:N ratio, which increases the persistence of soil cover over time. On the other hand, the leguminous crops, because they fix the atmospheric N, have high levels of N in the vegetal matter, and the vegetal remains generally have a low C:N ratio, with relatively fast decomposition, promoting small soil cover [65, 66], but provide significant amounts of N to the next crops. Simultaneous cultivation of leguminous, grassy, and oleaginous species has the potential to double the rate of addition of biomass to the soil, addition of nitrogen, and cycling of nutrients such as phosphorus, potassium, magnesium, and sulfur (**Table 4**) [61]. The studies have shown that the simultaneous cultivation of green manures in the Brazilian semiarid region can add a large amount of carbon and nutrients to the soil in agricultural systems, in a short period of time, not exceeding 70 days, during which period most species are in full bloom stage and are managed.

CC	DM	N	P	K	Mg	S
	Mg ha^{-1}			kg ha^{-1}		
PM 1	8.73 a	300.24 a	28.81 a	203.68 a	30.30 a	27.60 a
PM 2	8.51 a	268.11 b	30.32 a	214.55 a	31.25 a	30.90 a
EV	4.09 b	103.66 c	15.15 b	111.27 b	16.70 b	9.70 b
VC (%)	6.09	9.96	8.90	12.99	6.90	14.92

Adapted from [61]. The means followed by the same letter do not differ statistically from each other by Tukey test at the 5% probability level. CC—cover crop; M—Management; PM 1—plant mixture 1 (75% leguminous +25% grasses and oilseeds); PM 2—plant mixture 2 (25% leguminous +75% grasses and oilseeds); SV—spontaneous vegetation; NT—not tillage; T—tillage.

Table 4. Means of dry matter phytomass (DM), potassium (K), and sulfur (S) contents and nitrogen (N), phosphorus (P), potassium (K), magnesium (mg), and sulfur (S) accumulation of five cycles of plant mixtures crop and maintenance of spontaneous vegetation between rows of mango orchard.

Emphasizing on the importance of no-tillage and green manuring on soil protection mechanisms and residence time of carbon in the soil containing clay, studies have shown that protection mechanisms involving organo-mineral interactions are more important in carbon accumulation than occlusion within aggregates. However, in the Brazilian semiarid region, where most soils predominate sand fraction, both the interaction with minerals (organo-mineral) and physical protection are limited. In this case, carbon storage in the soil may depend on a fragile and continuous equilibrium of addition rate and decomposition that occurs naturally in these environments [67]. However, once the equilibrium has been broken down by means of crops and irrigation systems, there is a need to develop soil and crop management systems that allow the balance between rates of addition and decomposition that, at a minimum, maintain the levels similar to those found in soils under Caatinga. Therefore, the use of systems with a higher degree of complexity/diversity for both rainfed agriculture and irrigated agriculture can be an important strategy to promote low carbon agriculture, including the efficient management of water resources and salinization process and carbon and water footprints.

4.4. Water in the semiarid region

Only 3% of the total water in Brazil is in the semiarid region, with 78% located in the São Francisco and Parnaíba River basins. The temporal variability of the precipitations and the dominant geological characteristics, where there is predominance of shallow soils based on crystalline rocks and, consequently, low water changes between the river and the adjacent soil, results in the predominance of intermittent rivers and few perennial rivers. The semiarid region is a low volume region of river water flow [68]. The exploitation of groundwater is limited and presents problems due to the water that presents high content of salts and low flow wells (~1 m^3 h^{-1}), since over 80% of the crystalline region is about rocks [69]. However, the absence of rainfall is responsible for the insufficient supply of water in the region, but its poor distribution, associated with a high rate of evapotranspiration results in the phenomenon of drought and directly affects the population of the region. For this reason, the delimitation of the Brazilian semiarid zone is based on three technical criteria: average annual rainfall of less than 800 mm; index of aridity less than 0.5, calculated by the water balance that relates rainfall and potential evapotranspiration (I = P/ETP), between 1961 and 1990; drought risk (days with hydro citric acid/year greater than 60% per year), based on the period between 1970 and 1990.

Most of the population and rural properties of the Brazilian semiarid region depend on rainwater for human consumption and for agricultural and livestock production. Rain-dependent environments occupy the largest area. The capture and management of rainwater have been a popular technique developed by different peoples in different parts of the world, and there are thousands of people, especially in arid and semiarid regions [70]. The population throughout the history of coexistence with drought was developing different strategies to deal with human, animal, and primary food, fiber, and energy production [71].

In relation to rainwater harvesting in the Brazilian semiarid region, two main problems are highlighted: first, low rainfall utilization, mainly due to the use of large reservoirs, large reservoirs that concentrate water in large water mirrors that facilitate evaporation; second, storage and use of water by processes of higher points of drainage for the accumulation at lower points of the land. In its displacement to the storage site, water transports particles, contaminating it [72]. To handle the issues of capitation, storage, and water productivity, several techniques for harvesting rainwater were developed by Brazilian semiarid inhabitants to increase the availability of water for crop and livestock production. Among them, we highlight the reservoirs (Cisternas), surface dams (Barragens Subterrâneas), tank trench, water storage pits, small dam (Barraginha), and techniques for capturing rainwater in situ [73–75].

However, the irregularity of rainfall in the Brazilian semiarid region does not allow a production planning model dependent on precipitation during the crop development cycle. However, the integrated use of geotechnology, forecasting models, genetic improvement of plants, use of biotechnologies, and soil and water management strategies can boost water productivity. Thus, the efficient management of the water resource assumes great importance to mediate soil-plant-environment relations in a favorable way to compose a productive and sustainable system in the semiarid region, both for rain-dependent and irrigated environments.

4.5. Integrated crop-livestock-forest system

The Caatinga is rich in forage species in its three strata: herbaceous, shrub, and arboreal. Approximately 70% of the botanical species of the Caatinga take part significantly in the diet composition of the herds in the Brazilian semiarid region [76]. Facing the rational management of the Caatinga, agroecosystem models were developed so that farmers could have native or cultivated fodder throughout the year for their herds, increasing drought resilience and now the impacts of climate change.

The first researches identified the forage potential of native and exotic species. Among the native species are manicoba (*Manihot pseudoglaziovii* Pax & Hofman), manioc (*Manihot sculenta* Crantz), porcupine (*Manihot* sp), venom papaya (*Jakarta corumbensis* O. Kuntz), postumeira (*Gomphrena elegans* Mart. elegans), mandacaru without thorn (*Cereus hildemanianus* K Schum), camaratuba (*Cratylia argentea* desv. Kuntze), umbuzeiro (*Spondias tuberosa* Arr. Cam.), mororo (*Bauhinia* sp), and sage (*Mimosa caesalpinifolia* Benth). Among these exotic species, the most widely studied species are Buffel grass (*Cenchrus* spp.), Urochloa (*Urochloa mosambicensis*), forage palms (*Opuntia cus-indica* (L.) Mill., *Nopalea cochenillifera* Salm-Dick), Leucaena leucocephala (Lam), gliricidia (*Gliricidia sepium* (Jacq), and algaroba (Prosopis juli ora (SW) DC) [77, 78]. The agroecosystem design basis for the Brazilian semiarid region is the integration of adapted native or exotic elements, giving rise to models capable of increasing the resilience of the productive systems both in relation to the current edaphic climatic codes and in relation to the different scenarios of climate change.

One of the first agroecosystems developed for the semiarid region was called CBL, because it contemplates Caatinga, Buffel, and Leucaena subsystems. The Caatinga is grazed for 2–4 months. Buffel, as a water stress tolerant grass, is used during dry periods, and finally, Leucaena is a leguminous that complements feeding as a protein source, in the form of hay or silage. A second system developed, called Sistema Glória, proposes that in the rainy season, the herd be maintained under alternating grazing conditions in areas of cultivated grasses (Buffel, urochloa, pangolão, and aridus grass), as well as native annual cycle pastures; with predominance of marmalade grass (*Brachiaria plantaginea*) and several species of annual herbaceous leguminous, mainly of Phaseolus genera, Centrosema, and Stylosanthes. Both systems, in the periods of extreme drought, provide as a forage support the Indian Fig (*Opuntia ficus-indica* (L) P.Will) or native species as the xique-xique (*Pilosocereus gounellei*) and mandacaru (*Cereus jamacaru* DC) [79–83]. In general, the agroecosystem most used for semiarid region is composed of perennial woody species, associated with crops and pastures [79–81] denominated agrosilvopastoril system. The species composition may vary depending on the type of soil and rainfall regime. The implantation of complex, stable, sustainable models integrating elements of local biodiversity, arboreal, shrub, and herbaceous stratum is still a challenge for models of crop-livestock-forest integration adapted to semiarid conditions, to climate change scenarios, needing further research.

Agriculture and livestock are very important activities in the dryland economy. The typical agricultural and livestock systems are characterized by high grazing density, slash and burn practices, and irrigated monocultures. Consequently, soil degradation occurs due to unsustainable soil management, decreasing soil carbon stock and biodiversity. The soil carbon depletion is also associated with saline, water, and thermal stresses, typical in dryland regions. Climate change must be considered as a potentializer of stress and degradation factors. The environmental impacts of a warming climate in the semiarid region create challenges as well as opportunities.

The physical, chemical, and biological degradation process can be avoided and climatic resilience increased by improving science and technologies for low carbon agriculture, building sustainable agroecosystems. The challenge is to develop state policies, internalized by the population, that promote the sustainable and socially just development of Brazil, incorporating definitively science, technology, and innovation structures that guarantee the supply of water, energy, food, health, and culture through actions to mitigate and adapt to climate change. Adaptation and mitigation actions to climate change will be fundamental to guarantee human well-being and the continuity of life in its diversity on the planet, as we know it. Science and technologies for dryland are important to intelligent design and organic and adapted agroecosystems. Plant species tolerant to the saline, water, and thermal stress, no-tillage system associated with green manure, agroforestry, and water management are alternatives that can reduce GHG emissions, increase soil carbon sequestration, and mitigate the impact of climate change in the dryland, as well as to improve overall food security while making farmers more profitable and farms more profitable in Brazilian semiarid region. The physical, chemical, and biological degradation process can be avoided and climatic resilience increased by improving science and technologies to build sustainable agroecosystems.

Author details

Vanderlise Giongo[1]*, Alessandra Monteiro Salviano[1], Francislene Angelotti[1], Tatiana Taura[1], Luiz Fernando Carvalho Leite[2] and Tony Jarbas Ferreira Cunha[1]

*Address all correspondence to: vanderlise.giongo@embrapa.br

1 Embrapa Semiárido, Petrolina, PE, Brazil

2 Embrapa Meio-Norte, Teresina, PI, Brazil

References

[1] Guerry AD, Polasky S, Lubchenco J, Chaplin-Kramer R, Daily GC, Griffin R, Ruckelshaus M, Bateman IJ, Duraiappah A, Elmqvist T. Natural capital and ecosystem services informing decisions: From promise to practice. Proc Nat Acad Sci. 2015;**112**:7348-7355

[2] Le Quéré C, Moriarty R, Andrew RM, Canadel JG, Zeng N. Global carbon budget. Earth System Science Data. 2015;**7**:349-396

[3] Green JMH et al. Research priorities for managing the impacts and dependencies of business upon food, energy, water and the environment. Sustainability Science. 2017; **12**:319-331

[4] IPCC. Change 2014: Synthesis report. In: Pachauri RK, Meyer L, editors. Fifth assessment report. Geneva, Swizerland: IPCC; 2014. pp. 1-151

[5] Lal R. Societal value of soil carbon. Journal of Soil and Water Conservation. 2014;**6**:186-192

[6] Schulz K, Voigt K, Beusch C, Almeida Cortez JS, Kowarik I, Walz A, Cierjacks A. Grazing deteriorates the soil carbon stocks of Caatinga forest ecosystems in Brazil. Forest Ecology and Management. 2016;**367**:62-70

[7] MCT - Ministério Da Ciência, Tecnologia e Inovação. Estimativas anuais deemissões de gases de efeito estufa no Brasil, Brasilia. 2 ed.; 2014. pp. 1-101

[8] IPCC (Intergovernmental Panel on Climate Change). Climate Change 2013: The Physical Science Basis. Summary for Policymakers. Contribution of Working Group I to the Fifth Assessment Report of the Intergovernmental Panel on Climate Change. Genebra, Suiça: IPCC; 2013. 33 pp

[9] Elbehri A, Challinor A, Verchot L, et al. FAO-IPCC Expert Meeting on Climate Change, Land Use and Food Security: Final Meeting Report; 23-25, 2017; FAO HQ, Rome. FAO and IPCC; 2017

[10] UNFCCC. FCCC/CP/2015/L.9/Rev.1: Adoption of the Paris Agreement. Paris, France: UNFCCC; 2015. pp. 1-32

[11] UNFCCC. Intend Nationally Determined Contributions (INDCs). 2015. http://www4.unfccc.int/submissions/indc/Submission%20Pages/submissions.aspx

[12] Rogelj J, den Elzen M, Höhne N, et al. Paris agreement climate proposals need a boost to keep warming well below 2°C. Nature. 2016;**534**:631-639

[13] Brazil. Intended Nationally Determined Contribution: Towards achieving the objective of the United Nations Framework Convention on Climate Change. 2015. http://www.itamaraty.gov.br/images/ed desenvsust/BRASIL-iNDC-inglês.pdf

[14] Houghton RA. Contemporary C cycle. In: Holland H, Turekien K, editors. Treatize on Geochemistry, Oxford: International Soil and Water Conservation Research; 2014. pp. 399-435

[15] Ministério da Agricultura Pecuária e Abastecimento (Brasil). Desenvolvimento sustentável. [Internet]. 2013. Available from: http://www.agricultura.gov.br/desenvolvimento-sustentavel [Accessed: 2017 June 10]

[16] BRASIL MAPA (Ministério da Agricultura, Pecuária e Abastecimento). Plano Setorial de Mitigação e de Adaptação às Mudanças Climáticas para a Consolidação de uma Economia de Baixa Emissão de Carbono na Agricultura. Brasília, DF: MAPA; 2012

[17] Instituto Brasileiro de Geografia e Estatística. Censo 2010 [Internet]. Rio de Janeiro; 2011. Disponível em: http://www.censo2010.ibge.gov.br

[18] MMA. Plano de Ação para Prevenção e Controle ao Desmatamento na Caatinga (versão preliminar). Brasil: Departamento de Políticas para o Combate ao Desmatamento, Ministério do Meio Ambiente; 2010

[19] Brasil. Ministério da Integração Nacional. Relatório final do grupo de trabalho interministerial para redelimitação do Semi-Árido nordestino e do polígono das secas. Brasília, DF: MIN; MMA; MCT; 2005. 1 CD-ROM

[20] Salazar LF, Nobre CA, Oyama MD. Climate change consequences on the biome distribution in tropical South América. Geophysical Research Letters. 2007;**34**:1-6

[21] Marengo JA, Alves LM, Alvala RCS, Cunha AP, Brito S, Moraes OLL. Climatic characteristics of the 2010-2016 drought in the semiarid Northeast Brazil region. Anais da Academia Brasileira de Ciências. 2017;**89**:1-13

[22] Huang J, Yu H, Guan X, Wang G, Guo R. Accelerated dryland expansion under climate change. Nature Climate Change. 2016;**6**:166-172. DOI: http://dx.doi.org/10.1038/nclimate2837

[23] Safriel U, Adeel Z, Niemeijer D, Puigdefabregas J, White R, Lal R, Winslow M, Ziedler J, et al. Chapter 22: Dryland systems. In: Millennium Ecosystem Assessment—Ecosystems and Human Well-being. Washington, DC: World Resources Institute; 2005. pp. 623-662

[24] Silva PCG, Moura MSB, Kiill LHP, Brito LTL, Pereira LA, Sá IB, Correia RC, Teixeira AHC, Cunha TJF, Guimarães Filho C. Caracterização do Semiárido brasileiro: fatores naturais e humanos. In: Sá IB, Silva PCG, editores Semiárido brasileiro: pesquisa, desenvolvimento e inovação. Petrolina: Embrapa Semiárido; 2010. pp. 18-48

[25] Campos JNB. Secas e políticas públicas no semiárido: ideias, pensadores e períodos. Estud Av. 2014;**28**:65-68

[26] Campos JNB. Paradigms and public policies on drought in northeast Brasil: Historical perspective. Envir Manag. 2015;**55**:1052-1063

[27] Sampaio EVSB. Caracterização do bioma Caatinga. In: Gariglio MA, Sampaio EVSB, Cestaro LA, Kageyama PY, editors. Uso sustentável e conservação dos recursos florestais da caatinga. Brasília: Serviço Florestal Brasileiro; 2010

[28] Correia RC, Kill LHP, Moura MSB, Cunha TJF, Jesus Junior LA, de Araujo JLP. A região semiárida brasileira. In: Voltolini TV, editor. Petrolina: Embrapa Semiárido; 2011. pp. 21-48

[29] Campello BF. O uso da energia de Biomessa no Biome Caatinga. V Semana do Meio Ambiente, 3 a 5 de junho de 2008; Recife, PE. Fundação Joaquim Nabuco

[30] Sampaio EVSB, Costa TL. Estoques e fluxos de carbono no semi-árido nordestino: Estimativas preliminares. Revista Brasileira de Geografia Física. 2011;**6**:1275-1291

[31] Lapola DM, Martinelli LA, Peres CA, Ometto JPHB, Fereira ME, Nobre CA, Aguiar APD, Bustamante MMC, Cardoso MF, Costa MH, Joly CA, Leite CC, Moutinho P, Sampaio G, Strassburg BNN, Vieira IC. Pervasive transition of the Brazilian land-use system. Nature Climate Change. 2014;**4**:27-35

[32] da Silveira LM. Agricultura familiar no Semi-Árido brasileiro no contexto de mudanças climáticas globais. In: Angelotti F, Sá IB, Menezes EA, Pellegrino GQ, editors. Mudanças climáticas e desertificação no Semi-Árido brasileiro. Petrolina, PE: Embrapa Semi-Árido; 2009

[33] Angelotti F, Sa IB, Melo RF. Mudanças climáticas e desertificação no Semi-Árido brasileiro. In: Angelotti F, Sá IB, Menezes EA, Pellegrino GQ, editores. Mudanças climáticas e desertificação no Semi-Árido brasileiro. Petrolina: Embrapa Semi-Árido; 2009. pp. 41-49

[34] Acosta Salvatierra LH, Ladle RJ, Barbosa H, Correia RA, Malhado ACM. Protected areas buffer the Brazilian semi-arid biome from climate change. Biotropica. 2017, 2017;**49**:753-760

[35] Santini M, Caccamo G, Laurentini A, Noce S, Valentini RA. Multi-component GIS framework for desertification risk assesment by na integrated índex. Applied Geography. 2010;**30**:394-415

[36] Bisaro A, Kirk M, Zdruli P, Zimmermann W. Global drivers setting desertification research priorities: Insights froma stakeholder consultation forum. Land Degradation and Development. 2014;**25**:5-16

[37] Atkin OK, Sceurwater I, Ponts TL. Respiration as a percentage of daily photosynthesis in whole plants in homeostatic at moderate, but not high, growth temperatures. New Phytologist. 2007;**174**:367-380

[38] Campos JHBDaC, Silva MT, Silva VDePR. Impacto do aquecimento global no cultivo do feijão-caupi, no Estado da Paraíba. Rev. Bras. Eng. Agríc Ambiental. 2010;**14**: 396-404

[39] Deconto JG, coord. Aquecimento global e a nova geografia da produção agrícola no Brasil. Campinas: Embrapa Informática Agropecuária, Unicamp; 2008. 82 pp. il., color

[40] Moreira JN, Guimarães Filho C. Sistemas Tradicionais para Produção de Caprinos e Ovinos. In: Voltolini TV, editor. Produção de caprinos e ovinos no Semiárido. Petrolina: Embrapa Semiárido; 2011. pp. 49-68

[41] Santana M. Estoque de carbono e nitrogênio estoques de carbono e nitrogênio em solos do sertão Pernambuco sob diferentes usos. Dissertação (Mestrado em Tecnologias Energéticas e Nucleares). Recife: Universidade Federal de Pernambuco; 2015. 61 f

[42] Benko-Iseppon AM, Soares-Cavalcanti NM, Belarmino LC, Bezerra Neto JP, Amorim LLB, Fereira Neto JRC, Pandolfi V, Azevedo HMA, Silva RLO, Santos MG, Alves MV, Kido EA. Prospecção de Genes de Resistência à Seca e à Salinidade em Plantas Nativas e Cultivadas. Revista Brasileira de Geografia Física. 2011;**6**:1112-1134

[43] Sousa CAF, Cunha BAD, Martins PK, Molinari HBC, Kobayashi AK, Souza MT. *Nova* abordagem para a fenotipagem de plantas: conceitos, ferramentas e perspectivas. Revista Brasileira de Geografia Física. 2015;**8**:660-672

[44] Aidar ST, Morgante CV, Chaves ARM, Cruz JL, Costa Neto BP, Vitor AB, Martins DRPAS, Silva R, Cruz JL, Oliveira EJ. Physiological characteristics, total root and shoot production in accessions of *Manihot esculenta* under water stress condition. Revista Brasileira de Geografia Física. 2015;**8**:685-696

[45] Oliveira EJ, Aidar S de T, Morgante CV, Chaves AR de M, Cruz JL, Coelho Filho MA. Genetic parameters for drought-tolerance in cassava. Pesq Agrop Bras. 2015;**50**:233-241

[46] Araújo FP, Menezes EA, Santos CAF. Guandu Petrolina: uma boa opção para sua alimentação. Brasília, DF: Embrapa Informação Tecnológica; Petrolina, Embrapa Semi-Árido; 2007. 19 pp. il. (ABC da Agricultura Familiar, 14)

[47] Costa ND, Santos CAF, Queióz MA, Araújo HM, Oliveira VR, Mendonça JL, Candeia JA. Alfa São Francisco: variedade de cebola para cultivo de verão.2005. In: XLV Congresso Brasileiro de Olericultura; Fortaleza, Associação Brasileira Horticultura. CD-ROM

[48] Aidar ST, Chavea ARM, Fernandes-Junior PI, Oliveira MS, Costa Neto BP, Calsa Junior T, Morgante CV. Vegetative desiccation tolerance of Tripogon Spicatus (Poaceae) from the tropical semiarid region of northeastern Brazil. Functional Plant Biology. 2017;**44**:17066

[49] Amaral FCS, Tavares SRL. Alterações físico-químicas de um argissolo sob irrigação no perímetro irrigado Nilo coelho, Município de Petrolina-PE. Rio de Janeiro: Embrapa Solos; 2006. 46 pp. il .color. (Embrapa Solos. Boletim de pesquisa e desenvolvimento, 94)

[50] Masters DG, Benes SE, Norman HC. Biosaline agriculture for forage and livestock production. Agriculture, Ecosystems & Environment. 2007;**119**:234-248

[51] Porto ER, Araújo GGL. de Erva Sal (Atriplex nummularia). Petrolina, PE: Embrapa-CPATSA; 1999. 4 pp.il. (Embrapa Semi-Árido. Instruções Técnicas 22)

[52] Smith SE, Read DJ. Mycorrhizal Symbiosis. 3rd ed. London: Academic Press; 2008. 787 pp

[53] Kavamura VN, Santos SN, Silva JL, Parma MM, Ávila LA, Visconti A, Zucchi TD, Taketani RG, Andreote FD, Melo IS. Screening of Brazilian cacti rhizobacteria for plant growth promotion under drought. Microbiological Research. 2013;**168**:183-191

[54] Fernandes-Júnior PI, Aidar ST, Morgante CV, Gava CAT, Zilli JÉ, Souza LSB, Marinho RCN, Nóbrega RSA, Brasil MS, Seido SL, Martins LMV. The resurrection plant *Tripogon spicatus* (Poaceae) harbors a diversity of plant growth promoting bacteria in northeastern Brazilian Caatinga. Revista Brasileira de Ciência do Solo. 2015;**39**:993-1002

[55] Van Der Heidjen MGA, Horton TR. Socialism in soil? The importance of mycorrhizal fungal networks for facilitation in natural ecosystems. Journal of Ecology. 2009;**97**:1139-1150

[56] Suarez R, Wong A, Ramirez M, Barraza A, Orozcomdel C, Cevallo MA, Lara M, Hernandez G, Iturriaga G. Improvement of drought tolerance and grain yield in common bean by overexpressing trehalose-6-phosphate synthase in rhizobia. Molecular Plant-Microbe Interactions. 2008;**21**:958-966

[57] Neumann E, George E. The effect of arbuscular mycorrhizal root colonization on growth and nutriente uptake of two diferente cowpea (*Vigna unguiculata* [L.] Walp.) genotypes exposed to drought stress. Emirates Journal of Food and Agriculture. 2009;**21**:1-17

[58] Martins LMV, Xavier GR, Rangel FW, Ribeiro JRA, Neves MCP, Morgado LB, Rumjanek NG. Contribution of biological nitrogen fixation to cowpea: A strategy for improving grain yield in the semi-arid region of Brazil. Biology and Fertility of Soils. 2003;**38**:333-339

[59] Marinho RCN, Nóbrega RSA, Zilli JE, Xavier GR, Santos CAF, Aidar ST, Martins LMV, Fernandes Júnior PI. Field performance of new cowpea cultivars inoculated with efficient nitrogen-fixing rhizobial strains in the Brazilian semiarid. Pesquisa Agropecuária Brasileira. 2014;**49**:395-402

[60] Pereira Filho A, Giongo V, Cunha TJF, Texeira Filho J, Santos TL, Leite LFC. Low-carbon technology in brazilian semiarid ecosystems. In: Encyclopedia of Soil Science. 3rd ed. United Kingdom: Taylor & Francis; 2016. pp. 1-5. DOI: 10.1081/E-ESS3-120053711

[61] Brandão SS, Giongo V, Olszevski N, Mendes AMS. Coquetéis vegetais e sistemas de manejo alterando a qualidado do solo e produtividade da mangueira. Revista Brasileira de Geografia Física. 2017;**10**:1079-1089

[62] Pereira Filho A, Teixeira Filho J, Giongo V, Simões WL, Lal R. Nutrients dynamics in soil solution at the outset of notill implementation with the use of plant cocktails in Brazilian semi-arid. African Journal of Agricultural Research. 2016;**11**:234-246

[63] Giongo V, Mendes AMS, Santana MS, Costa ND, Yuri JE. Soil management systems for sustainable melon cropping in the submedian of the São Francisco Valley. Revista Caatinga. 2016;**29**:537-547

[64] Summers CF et al. Single season effects of mixed-species cover crops on tomato health (cultivar celebrity) in multi-state field trials. Applied Soil Ecology. 2014;**77**:51-58

[65] Perin A et al. Acúmulo e liberação de P, K, Ca e Mg em crotalária e milheto solteiros e consorciados. Revista Ceres Viçosa. 2010;**57**:274-281

[66] Perin A et al. Produção de fitomassa, acúmulo de nutrientes e fixação biológica de nitrogênio por adubos verdes em cultivo isolado e consorciado. Pesquisa Agropecuária Brasileira. 2004;**39**:35-40

[67] Conceição PC, Dieckow J, Bayer C. Combined role of no-tillage and cropping systems in soil carbon stocks and stabilization. Soil & Tillage Research. 2013;**129**:40-47

[68] Cirilo JA, Montenegro GL, Suzana, José Nilson C. A questão da água no Semiárido brasileiro. 2010. pp. 79-91

[69] Cirilo JA. Políticas públicas de recursos hídricos para o semi-árido brasileiro. Vol. 63. São Paulo: Universidade de São Paulo; 2008. pp. 61-82

[70] dos Santos DB, Medeiros SS, Brito LTL, Gnaldlinger J, Cohim E, Paz VPS, Gheyi HR, coord. Captação, manejo e uso da água da chuva. Campina Grande: INSA, ABCMAC; 2015. 440 pp

[71] Adham A, Riksen M, Ouessar M, Ritsema M. Identification of suitable sites for rainwater harvesting structuresin arid and semi-arid regions: A review. International Soil and Water Conservation Research. 2016, 2016;**4**:108-120

[72] Silva DF, Alcântara CR. Déficit Hídrico na Região Nordeste: Variabilidade Espaço [Temporal in the Northeast region: Space-time variability]. Cient. Exatas Tecnol., Londrina. Nov. 2009;**8**(1):45-51

[73] Silva AS, Porto ER, Gomes PCF. Seleção de áreas e construção de barreiros para uso em irrigação de salvação no Trópico Semi-Árido. Petrolina, PE: Embrapa-CPATSA; 1981. 43 pp. (EMBRA-PA-CPATSA. Circular Técnica, 3)

[74] Silva AS, Porto ER, Brito LTL, Monteiro MAR. Captação de água de chuva "in situ" I: Comparação de métodos da região semi-árida brasileira. In: EMBRAPA. Centro de Pesquisa Agropecuária do Trópico Semi-Árido. Captação de água de chuva "in situ": comparação de métodos e densidade de plantio. Petrolina, PE: EMBRAPA-CPATSA; 1989. pp. 5-24 (EMBRAPA-CPATSA, Boletim de Pesquisa, 35)

[75] Brito LTL, Silva DA, Cavalcanti NB, Anjos JB, Rêgo MM. Alternativa tecnológica para aumentar a disponibilidade de água no semi-árido. Revista Brasileira de Engenharia Agrícola e Ambiental. 1999;**3**:111-115

[76] Ferreira MA, Silva FM, Bispo SV, Azevedo M. Estratégias na suplementação de vacas leiteiras no semiárido do Brasil. Revista Brasileira de Zootecnia. 2009;**38**:322-329

[77] Araújo Filho JA, Carvalho FC. Sistema de produção agrossilvipastoril para o Semiárido Nordestino. In: Carvalho MM, Alvim MJ, Carneiro JC, editores. Sistemas agroflorestais pecuários: opções de sustentabilidade para áreas tropicais e subtropicais. Juiz de Fora: Embrapa Gado de Leite; 2001. pp. 102-110

[78] Voltolini TV, Neves ALA, Guimarães Filho C, Sa CO, Nogueira DM, Campeche DFB, Araújo GGL, Sa JL, Moreira JN, Veschi JLA, Santos RD, Moraes SA. Alternativas alimentares e sistemas de produção animal para o Semiárido brasileiro. In: Sá IB, GAMA PCG, editores. Semiárido brasileiro: pesquisa, desenvolvimento e inovação. Petrolina: Embrapa Semiárido; 2010. pp. 201-242

[79] Balbino LC, Cordeiro LAM, Porfirio-da-Silva V, Moraes A, Martinez GB, Alvarenga RC, KICHEL AN, Fontaneli RS, Santos HP, Franchini JC, Galerani PR. Evolução tecnológica e arranjos produtivos de sistemas de Integração lavoura-pecuária-floresta no Brasil. Pesquisa Agropecuária Brasileira. 2011;**46**:i-xii

[80] Balbino LC, Barcellos AO, Stone LF. Marco referencial: integração lavoura-pecuária-floresta (ILPF). Brasília, DF: Embrapa; 2011. 130 pp

[81] Rangel JHA, Muniz EN, Moraes SA, Souza SF, Amaral AJ, Pimentel JCM. Integração lavoura pecuária floresta na região nordeste do Brasil. Ciência Veterinária nos Trópicos. 2017;**19**:75-84

[82] Sá JL, Sá CO. Sistema Glória de produção de leite para o semi-árido. Petrolina: Embrapa Semiário; 2006. 4 pp. (Embrapa Semiárido. Circular técnica, 77)

[83] Guimarães Filho C, Soares JGG. Sistema CBL para produção de bovinos no semi-árido. Petrolina: Embrapa-CPATSA; 1999. 4 pp. (Embrapa Semiárido. Instruções técnica, 2)

Chapter 7

Pepper Crop under Climate Change: Grafting as an Environmental Friendly Strategy

Consuelo Penella and Angeles Calatayud

Additional information is available at the end of the chapter

http://dx.doi.org/10.5772/intechopen.72361

Abstract

Pepper is an extremely important vegetable worldwide in socio-economic terms. However, persistent land use, monoculture, and intensified production processes have led to soil diseases. This, along with abiotic stress, and mainly salinity of soil and waters, water stress, and suboptimal temperatures, can lead to physiological disorders emerging in peppers, e.g., cracking and Blossom end rot, which induce plant senescence, and lower not only in yields, but also in product quality. Salinity and water shortage are the two main environmental problems that crops face in the Mediterranean Region. One way of overcoming stresses from an ecological or integrated crop management viewpoint is to use grafted plants as an adaptation strategy. Initially, grafting technology has expanded in Solanaceae and Cucurbitacea species to overcome biotic stress. Nowadays, grafts are being used as several approaches to cushion the impact of climate change on agricultural systems. Furthermore, grafts allow desirable varieties by organoleptic or productivity traits, but they are sensitive to abiotic stress and can be grown under abiotic stress. As far as we know, very few studies on grafted pepper plants under abiotic stress are available.

Keywords: abiotic tolerance, drought, graft, pepper, salinity

1. Introduction

Peppers, chiles, capsicum, or no matter what other name they come under, are versatile crops included in most daily diets, especially in some areas more than others. Capsicum plants are topics crops that better grow in hotter zones [1]. They are eaten fresh, dehydrated and processed, and also as a spice. Given its vast versatility, peppers are being increasingly eaten, but also due to the fact that they are a major source of pro-vitamin A (carotene), E (α-tocopherol), and one of its main attributes is vitamin C (ascorbic acid). Mature pepper fruits are rich in carotenoids, compounds with anti-carcinogenic and antioxidant ability. Mature and immature fruits contain high contents of phenolics, especially flavonoids for which there are reports of antioxidant and other bioactive properties [2–4], and plenty of essential nutrients.

According to their culinary purposes and organoleptic features, pepper fruits are normally classified as two kinds. One is a bell pepper, which means a non-pungent, chunky sweet pepper kind, whereas chilli pepper refers to pungent chilli fruits [1]. Generally speaking, non-pungent peppers are more popular in the northern hemisphere, but more pungent chilli peppers are eaten more in tropic and subtropic areas [5].

Peppers grow in most countries on our planet, and they cover 1.93 million ha of crop-growing surface area. As a spice and vegetables, the world's pepper production has gone from over 12 million tons in 1993 to more than 31 million in 2013 over the past 20 years [6]. China is the largest pepper producer (almost 16 million tons) and is followed by Mexico (2.3 million), Turkey (2.2 million), and Indonesia (1.8 million) (**Figure 1**).

Peppers have adapted well to hot climates. The optimum seed germination temperature is 25–30°C. For fruit quality and growth purposes, areas with temperatures within the 21–29°C range are needed [7]. When temperature goes under 15°C or exceeds 32°C, growth can be retarded, and blossom end rot (BER), fruit-set ceases may emerge, with lower yields [8]. Generally speaking, commercial pepper varieties need friable, well-drained, sandy loam soil with pH of 6.5–7.5 for optimum production. Salt content in soil and irrigation water should be low. There are reports of a salinity resistance threshold of 1.5 dS m^{-1}, below which no effect on growth occurs, and a 14% drop in biomass production per additional 1 dS m^{-1} has been found [9]. Thresholds ranging from 0 to 2 dS m^{-1}, and slopes of salinity response curves that go from 8 to 15%, have been indicated for greenhouse peppers [10, 11]. Added organic matter increases the water-holding capacity and supplies minerals and nutrients. Peppers need high frequent soil fertility at the start of the growing cycle to supplement N. If water is lacking or excessive, flower abortion or further BER of fruits can be induced [12].

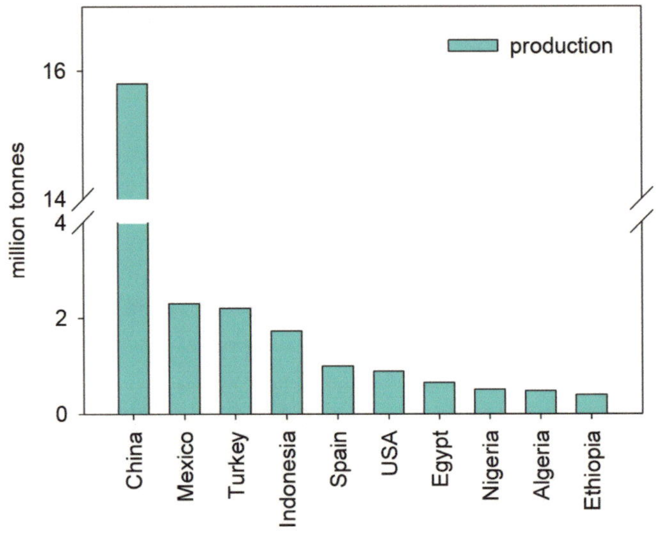

Figure 1. World production of chillies and peppers by country (million tons) [6].

1.1. Historical and botanical perspectives

The Solanaceae family is a complex that comprises at least 98 genera and as many as 2716 species, including Capsicum [13, 14]. This family also includes other major crop types, like potato, eggplant, tomato, and tobacco. "Capsicum" comes from a Greek-based derivate of the Latin "Kapto," which means "to bite," and refers to heat (pungency). Capsaicin, which is a volatile molecule, is also a very stable molecule that is responsible for the pungency normally linked with certain peppers [15]. Other pepper species are non-pungent because of a single mutation, which leads to the inability to generate capsaicinoids.

The genus Capsicum has been found in the central hemisphere and in South America ever since civilization began. It is likely to have evolved from an ancestral form in Bolivia-Peru. It formed part of human diet at approximately 7500 BC [16]. Peppers were completely unknown in Europe, Africa, and Asia before Columbus landed in the Americas. During his voyage, he came across a plant with fruit that resembled the pungency of black pepper, *Piper nigrum* L. The genus Capsicum, more commonly known as "pepper," "capsicum," "red chile," "bell pepper," "chilli pepper," "paprika," "tabasco," "cayenne," etc., contains up to around 40 species. The vast phenotypic variation comprises fruit shapes, colors, sizes, and plant habits [17]. Capsicum species, with barely any exceptions, are diploid (2n = 24, less frequently 2n = 26) with similar karyotypes [18, 19]. Wild and cultivated chillies possess morphological differences that can be easily discerned. The fruit of all wild chilli forms come as small, berry-like red fruits, and birds are attracted by their sizes and colors. *Capsicum annuum* L., *Capsicum chinense* Jacq., *Capsicum frutescens* L., *Capsicum baccatum* L. (C. var. pendulum), and *Capsicum pubescens* R & P are the five main cultivated or half-cultivated Capsicum species [1, 20, 21]. *C. chinense*, *C. annuum*, and *C. frutescens* form a closely linked group, also known as "annuum Complex" [22] which, according to some authors, are not differentiated species.

2. Main environmental problems to cultivate pepper plants

In the face of climate change, global food security demands increasing agricultural production on finite arable land that does not increase water use [23]. As the world's population is estimated to increase to about 9 billion by 2050, the World Food Summit on Food Security (2009) has set a target of a 70% global food production increase. Environmental stresses are the most limiting conditions for plant exploitation and horticultural productivity worldwide [24, 25]. The most limiting factors include temperature, water availability, light, salinity, pathogens, and metal ion concentrations. Many disorders and diseases can interfere with pepper production and its quality, which can be of biotic (living) and abiotic (non-living) origin.

2.1. Biotic stresses

Capsicum plants can be attacked by distinct pathogens. The most troublesome and important pests and diseases are: fungal diseases, like *Phytophthora capsici* (**Figure 2A** and **B**), *Rhizoctonia solani*, *Verticillium dahliae*, Fusarium spp., bacteria, e.g., *Xanthomonas campestris*, and powdery mildew (*Oidiopsis taurica* and *Leveillula taurica*), viruses (**Figure 2C**), like Tomato Spotted Wilt Virus (TSWV), Pepper Mottle Virus (PMV), Beet Curly Top Virus (BCTV), several Mosaic Virus

(AMV), (CMV), (TMV), nematodes, chiefly *Meloidogyne incognita*, and insects (**Figure 2D–I**), e.g., mites, termites, aphids, and thrips.

Biotic stresses can bring about physiological changes in pepper plants, e.g., ion-flux change, electrolyte leakage, activation of defensive responses, and hypersensitive cell death [26]. These effects can result in smaller yields and worse quality. One of the most hazardous biotic factors is soil diseases, especially for intensive farming, where soil-borne pathogens can build up if crop rotations are limited. The main injuries to roots of these soil pathogens include smaller foliar size, thin weak stems, wilting, depressed flowering, worse fruit quality, and shorter plant life spans [27]. Initial symptoms are quite visible on leaves when plant roots have been completely infected. Farmer's only feasible option is taking preventive measures, which involve soil treatments for the next crop season. As soil fumigation with methyl bromide (MB) is forbidden, other alternatives need to be taken [28]. Fumigants are an option, but vast amounts can be applied that might result in phytotoxicity [29, 30]. Furthermore, the long-term use of fumigants

Figure 2. (A) overview of a pepper field infected by *Phytophthora capsici* (courtesy of Juan José Tuset), (B) detail of roots infested with *P. capsici*, (C) virus (courtesy of J.I. Marsal), (D) overview of a pepper field with mites; leaf discoloring and defoliation, (E) buds parasitized by aphids (courtesy of A. Miguel), (F) flower detail with thrips (courtesy of J.I. Marsal), (G) pepper bitten by an insect, (H) stem affected by termites, and (I) details of aphids on a leaf (courtesy of J.I. Marsal).

may lead to changes in the microfauna of soil, which not always favors cultivated plants [31]. Steam treatment is not toxic and effectively kills pathogens, but is not an economically feasible option everywhere as it requires suitable steaming machinery, and also fuel and water [32]. Soil solarization is used frequently in countries with a warm climate [33], but soil must be covered for 4–6 weeks with hot periods to stop vegetable production. Another alternative is biological control, which involves selecting organisms based on their ability to control diseases, which can be used for aerial plagues.

Another possibility is plant biotic resistance. To enhance crop tolerance, many attempts have been made by traditional breeding programs. Although commercial success is limited given trait complexity, commercial cultivars with some tolerance are found. In the present-day, vast efforts have been made to genetically transform plants to improve their tolerance. Although some increased tolerance to pathogens has been reported in transgenic peppers [34, 35] other approaches to achieve resistance must be currently considered as genetic engineering means in plants has been poorly accepted by the public [36].

One way to reduce or avoid lost production is to graft sensitive plants onto robust rootstocks. Several *Capsicum* rootstocks, including breed lines, commercial cultivars, and wild accessions, can contribute adequate tolerance or resistance to *Phytophthora, Fusarium, Verticillium,* CMV, nematodes, etc. [37–39].

2.2. Abiotic stresses

During the growth cycle of peppers, as with other plants, many unfavorable environmental conditions can occur, such as salinity, drought, extreme temperatures, moisture, light, mineral deficiencies or toxicities, pH, and pollutants, which can all diminish plant yields [21, 40, 41]. Close to 82% of the potential crop yields is lost yearly from abiotic stress, and the quantity of available productive arable lands continues to drop worldwide, which forces farmers and farms to move to places with a higher abiotic stress potential [42].

In the Mediterranean Region, one of the most important abiotic stresses is salinity, which is usually present in both soil and water, as well water scarcity, but improving these environmental conditions through crop management is very difficult.

Some other abiotic stresses include: low temperature because it affects pepper vegetative development and reproduction as it disturbs how flower female organs function, and the amount of viable pollen grains per flower [43, 44]; high temperature and radiation promote stunted growth, a lower photosynthetic rate, increased respiration, and poor water and ion uptake [24, 26]. Therefore, using different shading screens is considered an alternative to overcome these problems [45, 46]. Likewise, heating is used to avoid chilling and frost injury, and cooling is employed to avoid high air temperatures [21].

2.2.1. Drought stress

Water scarcity is believed to be a major threat for the twenty-first century (UNESCO, 2012). During their life cycles, plants are subjected to periods of soil and atmospheric water deficit. Indeed, about only 15% of agricultural land is irrigated worldwide, but irrigated lands make up nearly 50% of the world's food production [47]. Drought, along with salinity, is one of the most important causes of low yields worldwide [48]. Adapted cultivars can improve the synchronization between crop water demand and soil supply. For all these reasons, we need to know plant responses to water scarcity, which are complex, and involve deleterious and/or adaptive changes [49].

As soil dries, its matric potential becomes more negative [50]. Plants can continue to absorb water only as long as their water potential (Ψw) is lower (more negative) than that of soil. The water potential is the total of both the solute potential (Ψs) and the turgor potential (Ψp): thus: $\Psi w = \Psi s + \Psi p$ [51]. In this way, one of the important pathways to enhance water stress tolerance is through osmotic adjustment, which maintains the leaf turgor required for stomatal opening, and to hence sustain photosynthesis and growth [52, 53]. Plants accumulate various types of compatible solutes, such as sugars, proline, glycinebetaine, or potassium [53, 54] to lower the osmotic potential and to absorb water. Basically, cells' accumulation of solutes is a process by which the water potential can lower without being linked to an accompanying reduction in turgor or a reduced cell volume.

Stomatal closure and reduced transpiration rates are prompt responses under drought stress because they lower the water potential of plant tissues. As a result, photosynthesis lowers, mediated by diminished CO_2 availability that is caused by: (a) diffusion limitations via the mesophyll and/or stomata [55], known as stomatal effects; (b) altered CO_2 fixation reactions, mediated by reduced Rubisco activity, known as non-stomatal [56]. With water stress, as energy accumulates in plants, which consume less light energy through photosynthetic carbon fixation, reactive oxygen species (ROS) generation increases [57, 58]. Accumulation of sorbitol, mannitol, and proline, and the formation of radical scavenging compounds, e.g., ascorbate, glutathione, and α-tocopherol, can help plants to cope with water stress [59]. Such compounds play a dual role as the non-enzymatic antioxidants needed by plants to counteract the inhibitory metabolic effects of the ROS generated under water stress [60], and also in stabilizing proteins and enzymes, and in protecting membrane integrity [61]. Besides these physiological responses, plants also undergo morphological changes [62], like stunted growth and, consequently, smaller yields.

Generally, pepper plants are sensitive to water deficit due to big leaf areas and higher stomata conductance [63–65]. In the pepper production industry, drought imposes huge reductions in crop yields and quality, with significant economic losses of up to 70% [64, 66, 67]. The two most critical moisture stress stages in peppers are the initial establishment of transplanted plants and the stage prior to blossoming [17]. Thus, reduced yields and smaller fruits are frequently recorded under moisture stress conditions. Moreover, this scenario limits the water applied to peppers during rapid growth periods to reduce final yields [68].

2.2.2. Salinity

Salinity can be disastrous because it can have many direct and indirect harmful effects. It inhibits seed germination, induces physiological dysfunctions and often kills non halophyte plants, even at low concentrations, and also limits agricultural development [69, 70]. Salinization transforms fertile and productive land into barren land, and often leads to habitat and biodiversity loss [71]. Salt accumulating in excessive amounts in cultivated soils is a common problem, especially under irrigated conditions, which threatens food production globally [72, 73]. The indiscriminate use of large quantities of chemical fertilizers and overexploitation of aquifers have dramatically multiplied the surface area affected by salinity [27]. Today to a greater or lesser extent, a third of all irrigated lands worldwide is affected by salinity [74], which means smaller yields.

Salt stress has two components that negatively affect plant growth: osmotic component and ionic component. A high salt concentration lowers the water potential in soil, and results in water stress in plants, known as the osmotic salinity component. The accumulation of given toxic ions represents the ionic component [75].

The relative degree of each salt effect caused by different salinity levels and its consequences on crop production are not clearly understood [67]. Saline soils induced by protected culture are complex and can include high concentrations of K^+, Na^+, Ca^{2+}, Mg^{2+}, SO_4^{2-}, NO_3^-, and Cl^-, which differ from the saline soils induced by seawater, in which NaCl is the most soluble and widespread salt [52, 76]. High Na^+ concentrations lower Ca^{2+} and K^+ uptakes, which leads to reduced stomatal conductance that results in lower CO_2 concentrations and, consequently, lower photosynthesis. High Cl^- concentrations cause chlorophyll degradation and reduce actual quantum yields of PSII electron transport [77].

Salinity causes membrane destabilization [78], nutrient imbalances [79] and irreversible harm to plant tissues and cells [80]. It is well-accepted that growth inhibition by salt stress is linked with alterations to the hydric relationships in plants as a result of osmotic effects with certain ionic consequences.

Salt tolerance mechanisms include: (i) salt exclusion: plants limit salt accumulation in tissues by inhibiting root uptake. Some salt transport restriction strategies to sensitive tissues or organs have also evolved [81]. Plants' ability to regulate the transport and uptake of salts depends on these mechanisms: root cells' selectivity of uptake; preferential loading of K^+ instead of Na^+ onto the xylem by stele cells; salts removed from the xylem in upper root parts, leaf sheaths, and the stem according to the exchange of both K^+ and Na^+; (ii) salt excretion: halophytes often have anatomical structures, like salt bladders and salt glands, that are designed to eliminate any excess salt ions from plants to their environment; and (iii) intracellular ion compartmentation. The sequestration of ions or salts into leaf and/or shoot vacuoles is typically attributed to dicotyledonous halophytes. Such accumulation depends on tonoplast Na^+/H^+ antiporters and vacuolar H^+-translocating transporters that are induced by saline environments [82]. One immediate salt stress effect is cell alkalinization, which is linked with the Na^+/H^+ antiporters activity of tonoplast vesicles [78]. Here different types of compatible organic solutes and potassium ions, like proline and soluble sugar, accumulate in the cytoplasm to avoid dehydration and to maintain the osmotic-ionic balance between both two compartments [83], and to also stabilize subcellular structures, e.g., proteins and membranes [52, 84]. It has been observed in tolerant salt plants after the initial loss of cellular turgor that plants are able to induce an osmotic adjustment to the lower external water potential by compartmentalizing toxic ions in the vacuole and then synthesizing compatible solutes in the cytoplasm [78].

Pepper, and *C. annuum* in particular, is highly susceptible to salt stress. Negative effects on yields stem from disturbances the following: membrane permeability, water channel activity, ion imbalance, poor total photosynthesis, and stomatal conductance, which modify the carbon balance required to maintain both productivity and growth [72, 85–87].

3. Main disorders related to abiotic stress in pepper plants

3.1. Blossom end rot

Blossom end rot (BER) is a serious disorder known to affect peppers that grow under different environmental stresses. BER symptoms are linked with membrane leakage of cell solutes, cell plasmolysis, and membrane breakdown [88–90]. Thus fruit surfaces display water-soaked symptoms, and the tissue at the distal fruit portion ends up becoming discolored and necrotic. BER causes premature ripening and enhances fruit softening, which result in small-sized fruits [91] (**Figure 3**). In internal fruit tissues, BER develops in the necrotic region of the parenchymal tissue surrounding young seeds, and also in the distal placenta [89]. It is predominantly viewed that the cause of BER is inadequate calcium translocation to the fruit tip for rapid fruit expansion, which takes place under conditions that favor rapid fruit growth, e.g., bright light and high temperature. Hence, cell integrity is impaired with consequent tissue disintegration [92]. Since Ca^{2+} is thought to play a key role, BER is termed a "calcium-related disorder" [93]. BER incidence is related to environmental factors, like high salinity, water scarcity, high temperature, and ammonia nutrition, which contribute to Ca^{2+} deficiency [91, 94, 95]. However, a close relationship between calcium levels and BER cannot always be demonstrated [90]. Lantos [96] has shown that applying calcium does not necessarily reduce the yield losses caused by calcium deficiency.

The influence of stress on BER which occurs in peppers is partly based on not only increased NAD(P)H oxidase (an oxygen radicals-generating enzyme) activity, but also on higher ROS production, e.g., superoxide radicals, hydroxyl radicals, and singlet oxygen (O_2) in fruit apoplasts [91, 92, 97]. ROS are known to trigger cell death, which is characterized by the progressive loss of membrane integrity to result in cytoplasm swelling, and also in the release of cellular constituents [98], including loss of Ca^{2+} ions, which may explain the lower Ca^{2+} concentrations found mainly in the apoplast [88]. A certain amount of stress, caused by either a single or an interaction of several environmental factors, like high relative humidity, pathogenic stem diseases, and dry or saline soils, may have a negative effect on calcium uptake [99], which does not always end in a corresponding degree of BER [90].

Figure 3. Overview of the pepper fruits affected by BER (right) and details of necrotic tissue (left).

Two phytohormone types appear to especially interfere with BER affection, and also in opposite directions: abscisic acid (ABA) and bioactive gibberellins (GAs). The antagonism action between vegetative growth and Ca^{2+} has been reported by Lyon et al. [100]. Low Ca^{2+} in the nutrient medium has been indicated to result in very extensive root systems, which suggests great GA activity. Accordingly, a low Ca^{2+} supply might have caused the high BER incidence more indirectly through enhanced GA activity [88]. ABA, as an antagonist to GAs, is known for reducing plant susceptibility to stress; e.g., by promoting Ca^{2+} transport to fruits. Applying ABA to highly stressed tomato plants has been recently demonstrated to alleviate BER symptoms [101].

From a practical point of view, GA-signaling can be reduced by, for example, root restriction [102], by applying growth-retarding chemicals, and also by ABA [103, 104].

Basically, BER development involves several steps: stress enhances ROS production; ROS leads to lipid peroxidation with greater membrane leakiness which, in turn, leads to the rapid vacuolation of parenchyma cells and to loss of ions, which includes water-soluble apoplastic Ca^{2+}. This situation is also aggravated when plants are grown vigorously, when GAs levels are high and when ABA is low. All these are typical BER symptoms [94]. Thus final Ca^{2+} deficiency can be considered a result, but not the cause, of only BER.

To control BER solutions, reducing susceptibility to stress and alleviating stress severity are necessary by: (i) proper selection of suited production sites. However, this is not always possible, and environmental conditions are unpredictable; (ii) improving management practices, e.g., shade or applying calcium fruit sprays. However, not enough evidence is available to recommend their use to manage BER; or spraying ABA, which remains unavailable as a commercial solution (no commercial formulation and side effects); (iii) breeding and selecting stress-resistant cultivars. Sadly, programs are slow and obtaining a variety that collects commercial fruit attributes and a robust radicular system is difficult; (iv) robust rootstocks inducing higher production in horticultural crops, which leads to a larger leaf area in grafted tomato plants [105], maintains a greater net CO_2 assimilation in grafted cucumber plants [106, 107], and has also shown a vigorous root system that increases the absorption of water and minerals in pepper-grafted plants [108]. Thus, grafting susceptible plants onto robust rootstocks to reduce their susceptibility to stress can reduce the fruits affected by BER, maintain water uptake, contribute to better plant nutrition; consequently, calcium deficiency can diminish [109–111].

3.2. Fruit cracking

Fruit cracking is yet another frequent physiological disorder that lowers marketable fruit yields, but it is not such a serious commercial problem as BER. The cracks in cracked fruits normally spread through the wall into the locule area because of repeated shrinkage. Such expansion weakens fruit cuticles [112]. Incidence is affected by environmental factors, mainly by varietal characteristics [113]. Several studies have demonstrated the importance of the environment in cuticle cracking development, like low night vapor pressure deficit [114], relative humidity [115], and temperature [116]. Fruits that display a wider expansion-shrinkage amplitude are often associated with severe cracking symptoms. The water status of fruits is a key factor to establish fruit cracking severity [21]. Some solutions can include those that minimize changes in their water status. Indeed, the same strategies used to combat BER can be adopted. Nonetheless, maintaining a consistent optimized growing environment is the best way to avoid fruit cracking.

4. Coping with abiotic constraints

The impact of both unpredicted climate change and climate variability on agricultural productivity is most likely to become a major constraint to achieve greater food production, which means that developing crop genotypes that withstand ambient stresses a major food security strategy. Hence, crop improvement innovations are needed [117]. They entail making furious efforts, especially by breeding companies that use conventional breeding programs. However, commercial success is extremely limited given the complex trait and practical selection tools are lacking; e.g., genetic markers have rendered these tasks inefficient and slow processes to date [84, 118, 119]. Combining suitable commercial fruit characteristics (quality and high production) and resistance to environmental factors is extremely difficult, especially when growing traditional varieties for their adaptation and traits quality since they are highly stress-sensitive [120, 121].

More recently, major efforts have been made to achieve genetic transformation [122–124]. Transferring a single gene or a few genes has led to claims of improved abiotic stress tolerance [125, 126]. However, the nature of genetically complex mechanisms of abiotic stress tolerance, and any potential detrimental side effects, makes this task most difficult [118, 127]. Lack of public acceptance of genetic engineering means that searching for other strategies to generate improved tolerances to abiotic stresses in plants is a priority [63, 128].

One environmental-friendly technique for avoiding or reducing loss in commercial yields caused by abiotic stress conditions is to graft susceptible commercial cultivars onto rootstocks that are capable of reducing the negative effect of external stress on shoots [25, 27, 129, 131]. Using grafted plants is an eco-friendly strategy that allows plants to overcome both soil-borne diseases and environmental stress [25, 110, 132].

4.1. Grafting

Grafting is defined as the natural or deliberate fusion of plant parts to establish vascular continuity among them [133], as well as the resulting genetically composite organism functions as a single plant [134]. The term scion denotes the shoot piece that stems from a donor plant that will be the grafted plant's canopy. The term rootstock indicates a plant that receives and fuses with the scion, and functions as the grafted plant's root system.

Despite vegetable grafting being an ancient practice, grafting did not become a common practice in ornamental and herbaceous vegetables before the twentieth century [135]. Cultivating grafted horticultural plants began in Korea and Japan toward the end of 1920s by grafting watermelon plants to squash rootstocks [136]. Ever since, this technique has been employed in watermelon, melon, cucumber, eggplant, pepper, tomato, and ornamental cactus and has exponentially increased. Grafting is also utilized for untypical fruit vegetables like artichoke [137, 138]. The advantages that vegetable grafting offers are attributed mainly to rootstocks' resistance to soil-borne diseases (fungus, nematodes, and bacterial wilt), and also to better vigor and stress tolerance. The problems related with banning methylbromide for soil fumigation purposes have led to increased vegetable grafting in the USA and Europe in recent years.

Micro- and tube-grafting and cleft approach are techniques that reliably combine pepper scions with compatible rootstocks, and the same can be stated of tomato and eggplant [139]. Recently, tube-grafting has become the most popular method type. It consists in cutting the growing rootstock tip at an angle of 45° below cotyledons and attaching it to the scion, which has been preciously cut at the same 45° angle above cotyledons, and then using a clip to fix the rootstock and scion (**Figure 4**).

Commercial varieties are not normally chosen to cope with abiotic stress. So, an interesting method to cope with these problems is to graft onto robust rootstocks.

Although grafting is a widespread eco-friendly technique applied in melon, tomato, or eggplant, it has been exploited less in peppers. This is basically because rootstock genotypes are lacking, which are simultaneously tolerant to biotic or abiotic stresses and can also improve commercial yields to amortize the extra costs incurred by grafting.

The main reason for grafting pepper is to improve plant vigor, disease tolerance, and uniformity, but very few commercial pepper rootstocks are available. This is because attention has been paid mainly to biotic stresses, and only the high-value pepper transplants utilized for protected cultivations are produced as grafted plants [39, 140, 141].

However, the abiotic stress incidence is very high, and increasing global climate change is forecast, while salinity and water stress are found frequently in areas where peppers are growing. It is necessary to perform several screenings to find Capsicum plants that tolerate abiotic stress so they can be used as rootstocks. In order to select the appropriate rootstocks,

Figure 4. Pepper seedling grafted by the tube-grating method.

searching for resistances in wild pepper types is crucial to amplify genetic diversity [142]. Currently, wild species of pepper from gene banks have been screened and phenotypically characterized as being tolerant to salinity and water stress under control conditions, and then used as rootstocks in the field, where abiotic stress problems occur, and productivity of grafted plants has been evaluated [132, 143, 144].

4.1.1. Grafting to cope with salt stress

One of the several approaches followed to cushion the impact of salinity is to graft plants onto tolerant rootstocks [10], and is a common agronomic practice in melon and tomato. Some works into these species have been conducted to elucidate the mechanisms that are involved in grafted plants' increased salinity tolerance. Such increased tolerance is generally associated with plants' capacity to retain or exclude, and/or accumulate toxic ions, Na^+ and Cl^- in rootstock roots. Hence, this action limits their transport to leaves instead of through the induction of antioxidant systems to the synthesis of osmotically active metabolites [35, 145]. Other authors have suggested that the rootstock's influence on the salt tolerance of scions is owing to stomatal functions (changes in stomatal regulation and water relations) being more efficient controlled. What this suggests is that grafting incisions could alter the hormonal signaling between shoots and roots [146]. In other cases, the re-establishment of ionic homeostasis has explained increased tolerance [124]. Yet in grafted plants, the mechanism of resistance against salinity displays a high degree of complexity in relation to specific scion/rootstock interactions [145, 147], and may vary among species. As far as we are aware, very few studies have been conducted into pepper to elucidate whether the salt tolerance conferred by rootstocks is due, or not, to retention and/or exclusion mechanisms, as in melon or tomato, because of them being better able to alleviate the toxic effects of salts or of other processes; e.g., water relations being maintained or antioxidant capacity being enhanced.

Salt tolerance among pepper genotypes may vary [72]. Maas [9] has indicated a salinity resistance threshold of 1.5 dS m^{-1}, and below which they found no effect on growth, but a 14% drop in biomass production for each additional 1 dS m^{-1}. Thresholds within the 0–2 dS m^{-1} range and slopes of salinity response curves that go from 8 to 15% have been reported for greenhouse peppers [10]. Another example is to use irrigation water of 4.4 dS m^{-1} [67], which resulted in reductions of 46% in the pepper dry biomass and of 25% in marketable pepper fruits. Guifrida et al. [109] have reported that stunted growth caused by salinity attenuates in pepper-grafted plants, compared with the non-grafted plants, is primarily associated with a low salt ions uptake. Therefore, these ions are present in the grafted plants at lower concentrations rather than leaf turgor being maintained by osmotic adjustments.

Different tolerance mechanisms to salt stress (NaCl 40–80 mM) were observed in our experiments using tolerant accessions (previously selected) like rootstocks and commercial "Adige" cultivar as a sensitive scion. Increased fruit yield under salinity when grafted onto accessions *Capsicum chinense* Jacq. "ECU-973" (code 12) and *Capsicum baccatum* L. var. pendulum "BOL-58" (code 14) was measured. Higher productivity under field conditions for these grafted plants was due to their ability to restrict Cl^- transport to leaves, and also to reduced Na^+ loading in leaves and roots, which thus favored K^+ (Ca^{2+} and Mg^{2+}) uptake and allowed a lower osmotic potential at less energy costs.

Such traits had a weak but negative impact on photosynthesis, nitrate reductase activity, and lipid peroxidation in the grafted scion leaves compared with ungrafted plants (Adige). Tolerance to salinity in these grafted plants was expressed to maintain scions' ion homeostasis, and can consequently improve crop yields [148, 149].

Nevertheless, by using *C. annuum* (code A25) as a tolerant rootstock, we also observed a larger amount of marketable fruit (+75%) and lower Blossom-end Root incidence (−31%) in commercial pepper cultivar Adige grafted onto A25 (A/A25) compared with ungrafted plants (**Figure 5A** and **B**), but the accumulation of Na^+ and Cl^- in leaves and roots was similar in grafted or ungrafted plants. Another tolerant salt mechanism was found in this grafted plant. Despite continued salt ions uptake, A/A25 plants' buffer capacity was not superseded as a testimony by unaffected biomass production and photosynthesis. The high Na^+ and Cl^-, accumulations and their likely compartmentalization in the apoplastic space and/or vacuole to preserve the cytosol from ionic toxic effects could well occur. Tolerance may be attributed to the ability to maintain shoot/root growth under salt stress, which has been related to A/A25 plants' ability to limit, or protect, loss of CO_2 assimilation (**Figure 6**) and sink activity in growing organs [149, 150].

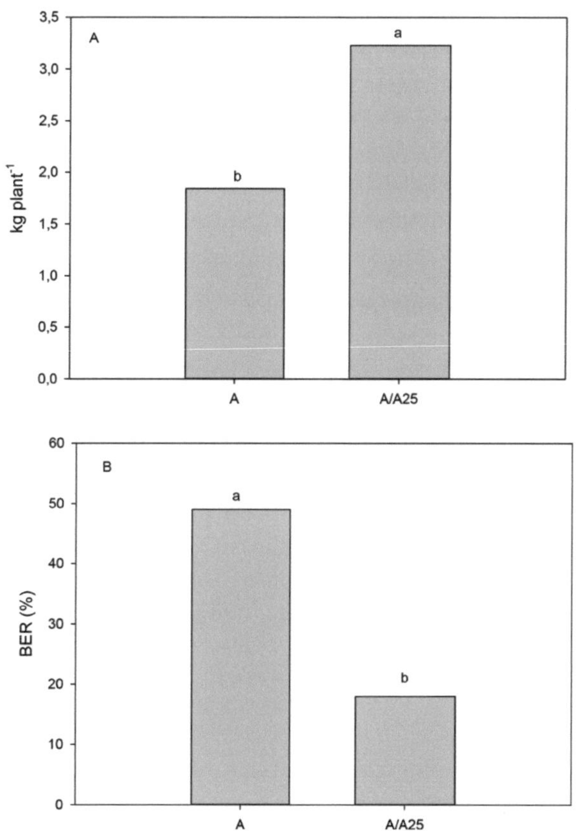

Figure 5. Marketable fruit yields (A) and the percentage of fruits affected by BER (B) under soil salinity and water conditions. Values are the mean of 50 replicates per cultivar Adige either ungrafted (A) or grafted onto the A25 genotype (A/A25). The different letters in each column denote significant differences at P < 0.05 according to the LSD test, and following a one-way ANOVA test by taking plant type as the variability factor.

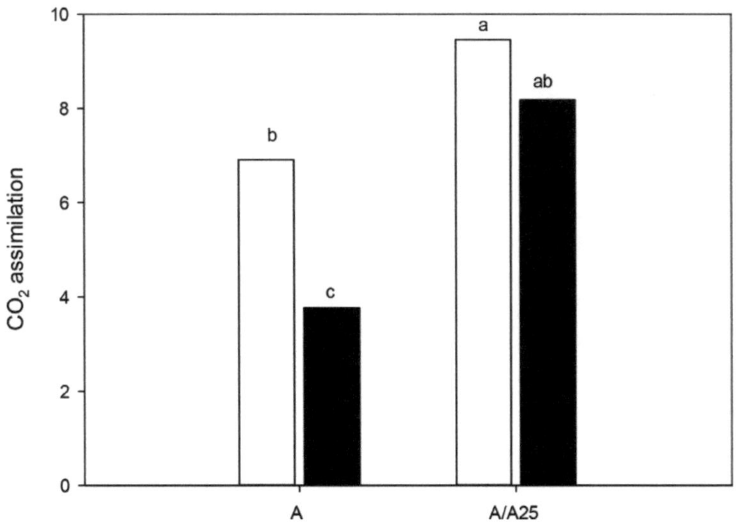

Figure 6. CO_2 assimilation (μmol CO_2 m^{-1} s^{-1}) of the cultivar Adige ungrafted (A) or grafted onto the A25 genotype (A/A25) under control (white bars) and salinity conditions (black bars). The values are the means of four replicates per genotype. The different letters in each column denote significant differences at $P < 0.05$ according to the LSD test, and following a two-way ANOVA test with plant type and NaCl treatment taken as the variability factor.

To conclude, grafting commercial varieties onto salt-tolerant rootstocks can be considered a valid strategy for ameliorating salt tolerance in peppers.

4.1.2. Grafting to overcome water stress

A novel perspective to enhance resistance to water stress is to use tolerant accessions as rootstocks for a given and desirable commercial cultivar. The interactions that take place among the graft, water stress, and vegetable plants have been studied mostly in cucumber, melon [151], and tomato [130, 152] by centering on the growth effects of grafting, and also on its physiological effects, and particularly on photosynthesis traits and hydric relationships [153]. Grafted plants usually show increased uptake of water and minerals compared to self-rooted plants as a result of the vigorous root system used as the rootstock [130, 154, 155]. Greater SOD and CAT activities, higher proline accumulation levels, and lower lipid peroxidation levels have been found in tobacco scions grafted onto drought-tolerant rootstocks [156]. Tomato grafted onto a drought-tolerant line has shown not only reduced growth, but also water conservation, as well as increased photosynthetic rates under mild drought conditions [152]. Similar results have been obtained by Liu et al. [157] using luffa as rootstocks when grafted with either its scion or cucumber.

However, reports on the physiological alterations of pepper after grafting and exposure to water stress are limited. Deep pepper root systems have been considered one of most important traits of tolerance. López-Marín et al. [158] have reported finding greater root growth in drought-tolerant grafted pepper plants (Hermino grafted onto Atlante) compared with scions (Herminio) ungrafted in an irrigation-deficit regime. The physiological tolerant mechanisms to overcome water stress in pepper-grafted plants are not well-known. The

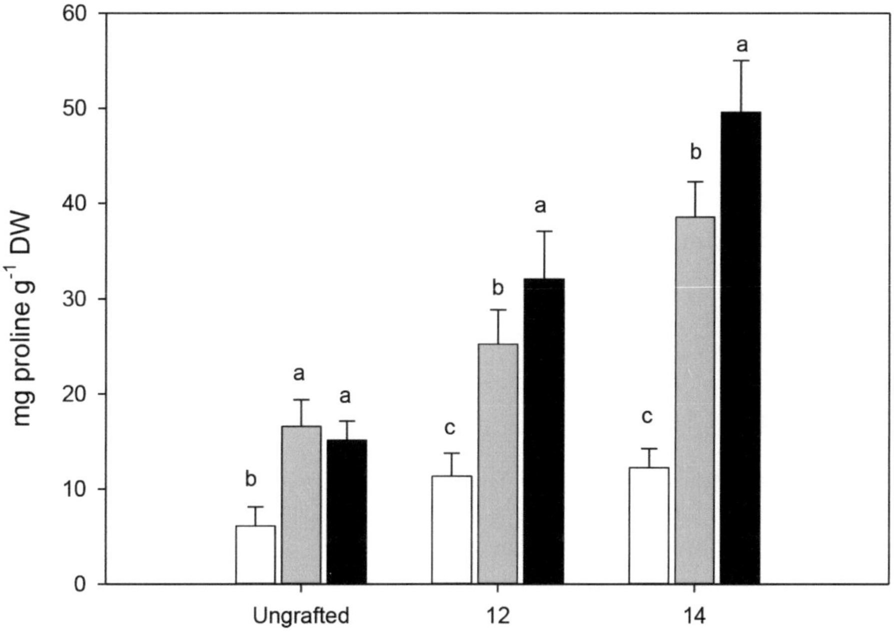

Figure 7. Changes in proline concentrations in leaves (mg proline g^{-1} DW) from the ungrafted pepper plants (cultivar "Verset") and the cultivar grafted onto accessions 12 and 14 after adding PEG at 0% (white bars), 3.5% (gray bars) and 7% (black bars) during a 14-day exposure period. Data are the mean values ± SE for n = 6. Within each plant combination, different letters indicate significant differences at $P < 0.05$ (LSD test).

effect of adding 3.5% and 7% PEG (polyethylene glycol) was examined for 14 days in two drought-tolerant rootstocks (codes 12 and 14, see Section 4.1.1) to identify the physiological traits responsible for the tolerance provided by rootstocks compared with ungrafted plants [159]. In grafted plants, we observed a higher proline level (**Figure 7**), along with a significant decrease in the osmotic potential, which reflected the lesser reduction in RWC. Enhanced osmotic adjustment may protect leaves from excessive dehydration. However, our results indicated that the water stress effect depended on the duration and intensity of the stress level, and also on the rootstock used.

Considering the overall results published about grafts, grafted plants can act as an efficient tool to mitigate abiotic stress in the climate change context and a tolerant rootstock that can make water and salt stress vanish on scions to reach greater productivity and fruit quality [149]. Nonetheless, the physiological and genetic mechanisms for abiotic tolerance in grafted plants, especially in peppers, are still unknown.

Acknowledgements

This work was financed by INIA (Spain) through Project RTA2013-00022-C02-01 and the European Regional Development Fund (ERDF).

Author details

Consuelo Penella and Angeles Calatayud*

*Address all correspondence to: calatayud_ang@gva.es

Departamento de Horticultural, Instituto Valenciano de Investigaciones Agrarias (IVIA), Moncada, Valencia, Spain

References

[1] De AK. Capsicum: The Genus Capsicum. London: CRC Press; 2003. 275 p

[2] Hervert-Hernández D, Sáyago-Ayerdi SG, Goñi I. Bioactive compounds of four hot pepper varieties (*Capsicum annuum* L.), antioxidant capacity, and intestinal bioaccessibility. Journal of Agricultural and Food Chemistry. 2010;**58**:3399-3406

[3] Mateos RM, Jiménez A, Román P, Romojaro F, Bacarizo S, Leterrier M, et al. Antioxidant systems from pepper (*Capsicum annuum* L.): Involvement in the response to temperature changes in ripe fruits. International Journal of Molecular Sciences. 2013;**14**:9556-9580

[4] Rodríguez-Burruezo A, Prohens J, Raigón MD, Nuez F. Variation for bioactive compounds in ají (*Capsicum baccatum* L.) and rocoto (*C. pubescens* R. & P.) and implications for breeding. Euphytica. 2009;**170**:169-181

[5] Cichewicz RH, Thorpe PA. The antimicrobial properties of chile peppers (*Capsicum* species) and their uses in Mayan medicine. Journal of Ethnopharmacology. 1996;**52**:61-70

[6] FAO (Food and Agriculture Organization of the United Nations). FAO Statistical Programme of Work. 2013. http://faostat3.fao.org/browse/Q/*/E

[7] Nonnecke IL. Crop salt tolerance. In: Tanji KK, editor. Agricultural Salinity Assessment and Management ASCE Manuals and Reports on Engineering Practices No 71. New York: American Society Civil Engineers-ASCE; 1989

[8] Knott JE, Deanon JR. Eggplant, tomato and pepper vegetable production in southeast asia. In: Los Banos, Laguna, Philippines: University of Philippines: Los Banos Press; 1967. pp. 99-109

[9] Maas EV. Crop salt tolerance. In: Tanji KK, editor. Agricultural Salinity Assessment and Management. ASCE Manuals and Reports on Engineering Practices No 71. New York: American Societay Civil Engineers-ASCE; 1976

[10] Chartzoulakis K, Klapaki G. Response of two greenhouse pepper hybrids to NaCl salinity during different growth stages. Scientia Horticulturae. 2000;**86**:247-260

[11] Navarro JM, Garrido C, Martínez V, Carvajal M. Water relations and xylem transport of nutrients in pepper plants grown under two different salts stress regimes. Plant Growth Regulation. 2003;**41**:237-245

[12] Maroto JV, Borrego JVM. Elementos de horticultura general. Mundi-Prensa Libros; 2008. 481 p

[13] Hunziker AT. In: Verlag G, editor. Genera Solanacearum. The Genera of Solanaceae Illustrated, Arranged According to a New System. Liechtenstein: Ruggel; 2001. 561 p

[14] Olmstead RG, Sweere JA, Spangler RE, Bohs L, Palmer JD. Phylogeny and provisional classification of the Solanaceae based on chloroplast DNA. In: Nee M, Symon DE, Lester RN, Jessop JP, editors. Solanaceae IV: Advances in Biology and Utilization. New, UK: Royal Botanic Gardens; 1999. pp. 111-137

[15] Heiser CB, Pickersgill B. Names for the cultivated Capsicum species. Taxon. 1969;**18**: 277-283

[16] MacNeish RS. The food-gathering and incipient agriculture stafe of prehistoric middle america. In: Wauchope R, editor. Natural Environments and Early Cultures. Handbook. Austin: University of Texas Press; 1964. p. 413-426

[17] Bosland PW, Votava EJ. Peppers: Vegetable and spice capsicums. In: Atherton J, Rees A. (eds.). Crop Production Science in Horticulture Series. New York: CABI Publishing; 2003. 204 p

[18] Lippert LF, Smith PG, Bergh BO. Cytogenetics of the vegetable crops. Garden pepper, capsicum sp. The Botanical Review. 1966;**32**:24-55

[19] Moscone EA, Lambrou M, Hunziker AT, Ehrendorfer F. C-banded karyotypes in *Capsicum* (Solanaceae). Plant Systematics and Evolution. 1993;**186**:213-229

[20] Macrae R. Encyclopedia of Food Science, Food Technology and Nutrition (Peppers and Chilies). Academic Press; 1993. 3496-3505 p

[21] Russo V. Peppers: Botany, Production and Uses. London: CAB International; 2012. 280 p

[22] Nuez F, Gil R, Costa J. El cultivo de pimientos, chiles y ajies. Mundi-Prensa; 1996. 607 p

[23] Davis KF, Rulli MC, D'Odorico P. The global land rush and climate change. Earth's Future. 2015

[24] Nilwik HJM. Growth analysis of sweet pepper (*Capsicum annuum* L.): 1. The influence of irradiance and temperature under glasshouse conditions in winter. Annals of Botany. 1981;**48**:129-136

[25] Schwarz D, Rouphael Y, Colla G, Venema JH. Grafting as a tool to improve tolerance of vegetables to abiotic stresses: Thermal stress, water stress and organic pollutants. Scientia Horticulturae. 2010;**127**:162-171

[26] Lee S, Choi D. Comparative transcriptome analysis of pepper (*Capsicum annuum*) revealed common regulons in multiple stress conditions and hormone treatments. Plant Cell Reports. 2013;**32**:1351-1359

[27] Rivero R, Ruiz J, Romero L. Role of grafting in horticultural plants under stress conditions. Food, Agriculture & Environment. 2003;**1**:70-74

[28] Batchelor TA, Miller MK. The role of the Montreal Protocol in reducing quarantine and preshipment uses of methyl bromide. In: Proceedings of the 8th International Conference on Controlled Atmosphere and Fumigation in Stored Products; Chengdu. 2008. p. 88-93

[29] Giannakou IO, Sidiropoulos A, Prophetou-Athanasiadou D. Chemical alternatives to methyl bromide for the control of root-knot nematodes in greenhouses. Pest Management Science. 2002;**58**:290-296

[30] Noe JP. Crop-and nematode-management systems. In: Barker KR, Perderson GA, Windham GL, editors. Plant Nematode Interactions. Madison: ASA, CSSA, SSSA Press; 1998

[31] Hague NGM, Gowen SR. Chemical control of nematodes. In: Brown RH, Kerry BR, editors. Principles and Practice of Nematode Control in Crops. Academic Press; 2014. pp. 131-178

[32] FAO (Food and Agriculture Organization of the United Nations). FAO Statistical Programme of Work. 2015

[33] Gordh G, McKirdy S. The Handbook of Plant Biosecurity: Principles and Practices for the Identification, Containment and Control of Organisms that Threaten Agriculture and the Environment Globally. Vol. 19. Springer Science & Business Media; 2014. 738 p

[34] Arthikala M-K, Sánchez-López R, Nava N, Santana O, Cárdenas L, Quinto C. RbohB, a *Phaseolus vulgaris* NADPH oxidase gene, enhances symbiosome number, bacteroid size, and nitrogen fixation in nodules and impairs mycorrhizal colonization. New Phytologist; 2014;**202**:886-900

[35] Watson RR, Preedy VR. Genetically Modified Organisms in Food: Production. Safety: Regulation and Public Health. Elsevier Science; 2015. 516 p

[36] Estañ MT, Martinez-Rodriguez MM, Perez-Alfocea F, Flowers TJ, Bolarin MC. Grafting raises the salt tolerance of tomato through limiting the transport of sodium and chloride to the shoot. Journal of Experimental Botany. 2005;**56**:703-712

[37] Kokalis-Burelle N, Bausher MG, Rosskopf EN. Greenhouse evaluation of Capsicum rootstocks for management of Meloidogyne incognita on grafted bell pepper. Nematropica. 2009;**39**:121-132

[38] Morra L, Bilotto M. Evaluation of new rootstocks for resistance to soil-borne pathogens and productive behaviour of pepper (*Capsicum annuum* L.). The Journal of Horticultural Science and Biotechnology. 2006;**81**:518-524

[39] Oka Y, Offenbach R, Pivonia S. Pepper rootstock graft compatibility and response to *Meloidogyne javanica* and *M. incognita*. Journal of Nematology. 2004;**36**:137-141

[40] Ashraf M. Genetic variation for salinity tolerance in spring wheat. Hereditas. 2004; **120**:99-104

[41] Munns R, James RA, Läuchli A. Approaches to increasing the salt tolerance of wheat and other cereals. Journal of Experimental Botany. 2006;**57**:1025-1043

[42] Hirt H, Shinozaki K. Plant Responses to Abiotic Stress. Springer Science & Business Media; 2004. 300 p

[43] Pressman E, Shaked R, Firon N. Exposing pepper plants to high day temperatures prevents the adverse low night temperature symptoms. Physiologia Plantarum. 2006; **126**:618-626

[44] Shaked R, Rosenfeld K, Pressman E. The effect of low night temperatures on carbohydrates metabolism in developing pollen grains of pepper in relation to their number and functioning. Scientia Horticulturae. 2004;**102**:29-36

[45] Ilahy R, Thouraya R, Tlili I, Jebari H. Effect of different shading levels on growth and yield parameters of a hot pepper (*Capsicum annuum* L .) Cultivar `Beldi` grown in Tunisia. Food. 2013;**7**:31-35

[46] López-Marín J, González A, Pérez-Alfocea F, Egea-Gilabert C, Fernández JA. Grafting is an efficient alternative to shading screens to alleviate thermal stress in greenhouse-grown sweet pepper. Scientia Horticulturae. 2013;**149**:39-46

[47] Tilman D, Cassman KG, Matson PA, Naylor R, Polasky S. Agricultural sustainability and intensive production practices. Nature. 2002;**418**:671-677

[48] Bodner G, Nakhforoosh A, Kaul H-P. Management of crop water under drought: A review. Agronomy for Sustainable Development. 2015;**35**:401-442

[49] Chaves MM, Pereira JS, Maroco J, Rodrigues ML, Ricardo CPP, Osorio ML, et al. How plants cope with water stress in the field. Photosynthesis and growth. Annals of Botany. 2002;**89**:907-916

[50] Taiz L, Zeiger E. Plant Physiology [Internet]. Sinauer Associates; 2010. 782 p

[51] Kramer PJ, Boyer JS. Water Relations of Plants and Soils [Internet]. Academic Press; 1995. 495 p

[52] Huang Y, Bie Z, He S, Hua B, Zhen A, Liu Z. Improving cucumber tolerance to major nutrients induced salinity by grafting onto *Cucurbita ficifolia*. Environmental and Experimental Botany. 2010;**69**:32-38

[53] Nio SA, Cawthray GR, Wade LJ, Colmer TD. Pattern of solutes accumulated during leaf osmotic adjustment as related to duration of water deficit for wheat at the reproductive stage. Plant Physiology and Biochemistry. 2011;**49**:1126-1137

[54] Morgan J. Osmotic components and properties associated with genotypic differences in osmoregulation in wheat. Australian Journal of Plant Physiology. 1992;**19**:67-76

[55] Flexas J, Diaz-Espejo A, Berry JA, Cifre J. Of leakage in IRGA's leaf chambers of open gas exchange systems: Quantification and its. Journal of Experimental Botany. 2007;**58**:1533-1543

[56] Lawlor DW, Cornic G. Photosynthetic carbon assimilation and associated metabolism in relation to water deficits in higher plants. Plant, Cell & Environment. 2002;**25**: 275-294

[57] Asada K. Production and scavenging of reactive oxygen species in chloroplasts and their functions. Plant Physiology. 2006;**14**:391-396

[58] Smirnoff N. The role of active oxygen in the response of plants to water deficit and desiccation. The New Phytologist. 1993;**125**:27-58

[59] Yordanov I, Velikova V, Tsonev T. Plant responses to drought and stress tolerance. Bulgarian Journal of Plant Physiology. 2003;**38**:187-206

[60] Gill SS, Tuteja N. Reactive oxygen species and antioxidant machinery in abiotic stress tolerance in crop plants. Plant Physiology and Biochemistry. 2010;**48**:909-930

[61] Patade VY, Bhargava S, Suprasanna P. Halopriming mediated salt and iso-osmotic PEG stress tolerance and, gene expression profiling in sugarcane (*Saccharum officinarum* L.). Molecular Biology Reports. 2012;**39**:9563-9572

[62] Vassileva V, Demirevska K, Simova-Stoilova L, Petrova T, Tsenov N, Feller U. Long-term field drought affects leaf protein pattern and chloroplast ultrastructure of winter wheat in a cultivar-specific manner. Journal of Agronomy and Crop Science. 2012;**198**:104-117

[63] Campos H, Trejo C, Peña-Valdivia CB, García-Nava R, Conde-Martínez FV, Cruz-Ortega MR. Stomatal and non-stomatal limitations of bell pepper (*Capsicum annuum* L.) plants under water stress and re-watering: Delayed restoration of photosynthesis during recovery. Environmental and Experimental Botany. 2014;**98**:56-64

[64] Delfine S, Alvino A, Loreto F, Centritto M, Santarelli G. Effects of water stress on the yield and photosynthesis of field-grown sweet pepper (*Capsicum annuum* L.). In: Proc Third Int Symp Irrig Hortic Crop. Vols 1-2. 2000. pp. 223-229, 942

[65] Gonzalez-Dugo V, Durand J-L, Gastal F. Water deficit and nitrogen nutrition of crops. A review. Agronomy for Sustainable Development. 2010;**30**:529-544

[66] Fernández MD, Gallardo M, Orgaz F, Thompson F, Fereres E. Water use and production of a greenhouse pepper crop under optimum and limited water supply. The Journal of Horticultural Science and Biotechnology. 2005;**80**:87-96

[67] Pascale S, De Ruggiero C, Barbieri G. Physiological responses of pepper to salinity and drought. Journal of the American Society for Horticultural Science. 2003;**128**:48-54

[68] Beese F, Horton R, Wierenga PJ. Growth and yield response of Chile pepper to trickle Irrigation1. Agronomy Journal. 1982;**74**:556

[69] Bartels D, Sunkar R. Drought and salt tolerance in plants. Critical Reviews in Plant Sciences. 2005;**24**:23-58

[70] Shannon MC. Adaptation of plants to salinity. Advances in Agronomy. 1997;**60**:75-120

[71] Ghassemi E, Jakeman A, HA N. Salinization of Land and Water Resources. Wallingford Oxon: CABI; 1995

[72] Aktas H, Abak K, Cakmak I. Genotypic variation in the response of pepper to salinity. Scientia Horticulturae (Amsterdam). 2006;**110**:260-266

[73] Zeng L, Lesch SM, Grieve CM. Rice growth and yield respond to changes in water depth and salinity stress. Agricultural Water Management. 2003;**59**:67-75

[74] Pasternak D. Salt tolerance and crop production—A comprehensive approach. Annual Review of Phytopathology. 1987;**25**:271-291

[75] Greenway H, Munns R. Mechanisms of salt tolerance in nonhalophytes. Annual Review of Plant Physiology. 1980;**31**:149-190

[76] Luo Q, Yu B, Liu Y. Differential sensitivity to chloride and sodium ions in seedlings of glycine max and *G . soja* under NaCl stress. Journal of Plant Physiology. 2005;**169**:1003-1012

[77] Mitra GN. Regulation of Nutrient Uptake by Plants: A Biochemical and Molecular Approach. Springer; 2015. 195 p

[78] Hasegawa PM, Bressan R, Zhu J-K, Bonhert HJ. Plant cellular and molecular responses to high salinity. Annual Review of Plant Physiology and Plant Molecular Biology. 2000;**51**:463-499

[79] Munns R. Physiological processes limiting plant growth in saline soils: Some dogmas and hypotheses. Plant, Cell & Environment. 1993;**16**:15-24

[80] Meyer RF, Boyer JS. Osmoregulation, solute distribution, and growth in soybean seedlings having low water potentials. Planta. 1981;**151**:482-489

[81] Munns R. Avenues for increasing salt tolerance of crops. Plant Nutrition. 2002:370-371

[82] Barkla BJ, Pantoja O. Physiology of ion transport across the tonoplast of higher plants. Annual Review of Plant Physiology and Plant Molecular Biology. 1996;**47**:159-184

[83] Munns R, Tester M. Mechanisms of salinity tolerance. Annual Review of Plant Biology. 2008;**59**:651-681

[84] Ashraf M, Foolad MR. Roles of glycine betaine and proline in improving plant abiotic stress resistance. Environmental and Experimental Botany. 2007;**59**:206-216

[85] Kurunc A, Unlukara A, Cemek B. Salinity and drought affect yield response of bell pepper similarly. Acta Agriculturae Scandinavica. Section B. Soil and Plant Science. 2011;**61**:514-522

[86] Piñero MC, Houdusse F, Garcia-Mina JM, Garnica M, Del Amor FM. Regulation of hormonal responses of sweet pepper as affected by salinity and elevated CO_2 concentration. Physiologia Plantarum. 2014;**151**:375-389

[87] Carvajal M, Martínez V, Alcaraz CF. Physiological function of water channels as affected by salinity in roots of paprika pepper. Physiologia Plantarum. 1999;**105**:95-101

[88] de Freitas ST, Jiang CZ, Mitcham EJ. Mechanisms involved in calcium deficiency development in tomato fruit in response to gibberellins. Journal of Plant Growth Regulation. 2012;**31**:221-234

[89] Ho LC, White PJ. A cellular hypothesis for the induction of blossom-end rot in tomato fruit. Annals of Botany. 2005;**95**:571-581

[90] Saure MC. Blossom-end rot of tomato (*Lycopersicon esculentum* mill.) — A calcium- or a stress-related disorder? Scientia Horticulturae. 2001;**90**:193-208

[91] Aktas H, Karni L, Chang DC, Turhan E, Bar-Tal A, Aloni B. The suppression of salinity-associated oxygen radicals production, in pepper (*Capsicum annuum*) fruit, by manganese, zinc and calcium in relation to its sensitivity to blossom-end rot. Physiologia Plantarum. 2005;**123**:67-74

[92] Turhan E, Aktas H, Deventurero G, Karni L, Bar-Tal A, Aloni B. Blossom-end rot is associated with impairment of sugar metabolism and growth of pepper (*Capsicum annuum* L.) fruits. The Journal of Horticultural Science and Biotechnology. 2006;**81**:921-927

[93] Lc H, Belda R, Brown M, Andrews J, Adams P. Uptake and transport of calcium and the possible causes of blossom-end rot in tomato. Journal of Experimental Botany. 1993;**44**:509-518

[94] Saure MC. Why calcium deficiency is not the cause of blossom-end rot in tomato and pepper fruit—A reappraisal. Scientia Horticulturae. 2014;**174**:151-154

[95] Taylor MD, Locascio SJ, Alligood MR. Blossom-end rot incidence of tomato as affected by irrigation quantity, calcium source, and reduced potassium. Hortscience. 2004;**39**:1110-1115

[96] Lantos F. A kalcium növényélettani szerepének, jelentő- ségének vizsgálata a paprikatermesztésben, hiánytüneteinek visszaszorítása növénynemesítői eljárásokkal. [Study of the physiological role of calcium in paprika (*Capsicum annuum*) production]. Agrár- és Vidékfejlesztési Szeml. 2007;**2**:125-129

[97] Mestre TC, Garcia-Sanchez F, Rubio F, Martinez V, Rivero RM. Glutathione homeostasis as an important and novel factor controlling blossom-end rot development in calcium-deficient tomato fruits. Journal of Plant Physiology. 2012;**169**:1719-1727

[98] Van Breusegem F, Dat JF. Reactive oxygen species in plant cell death. Plant Physiology. 2006;**141**:384-390

[99] Marcelis LFM, Ho LC. Blossom-end rot in relation to growth rate and calcium content in fruits of sweet pepper (*Capsicum annuum* L.). Journal of Experimental Botany. 1999;**50**:357-363

[100] Lyon CB, Beeson KC, Barrentine M. Macro-element nutrition of the tomato plant as correlated with fruitfulness and occurence of blossom-end rot. Bot Gazzete. 1942;**103**: 651-667

[101] Tonetto De Freitas S, McElrone AJ, Shackel KA, Mitcham EJ. Calcium partitioning and allocation and blossom-end rot development in tomato plants in response to whole-plant and fruit-specific abscisic acid treatments. Journal of Experimental Botany. 2014;**65**:235-247

[102] Karni L, Aloni B, Bar-Tal A, Moreshet S, Keinan M, Yao C. The effect of root restriction on the incidence of blossom-end rot in bell pepper (*Capsicum annuum* L.). The Journal of Horticultural Science and Biotechnology. 2000;**75**:364-369

[103] Lurie S, Handros A, Fallik E, Shapira R. Reversible inhibition of tomato fruit gene expression at high temperature (effects on tomato fruit ripening). Plant Physiology. 1996;**110**:1207-1214

[104] Saito S, Hirai N, Matsumoto C, Ohigashi H, Ohta D, Sakata K, et al. Arabidopsis CYP707As encode (+)-abscisic acid 8′-hydroxylase, a key enzyme in the oxidative catabolism of abscisic acid. Plant Physiology. 2004;**134**:1439-1449

[105] Albacete A, Martínez-Andújar C, Ghanem ME, Acosta M, Sánchez-Bravo J, Asins MJ, et al. Rootstock-mediated changes in xylem ionic and hormonal status are correlated with delayed leaf senescence, and increased leaf area and crop productivity in salinized tomato. Plant, Cell & Environment. 2009;**32**:928-938

[106] Colla G, Suarez CMC, Cardarelli M, Rouphael Y. Improving nitrogen use efficiency in melon by grafting. Hortscience. 2010;**45**:559-565

[107] Davis AR, Perkins-Veazie P, Sakata Y, López-Galarza S, Maroto JV, Lee S-G, et al. Cucurbit Grafting. Critical Reviews in Plant Sciences. 2008;**27**:50-74

[108] Leal-Fernández C, Godoy-Hernández H, Núñez-Colín CA, Anaya-López JL, Villalobos-Reyes S, Castellanos JZ. Morphological response and fruit yield of sweet pepper (*Capsicum annuum* L.) grafted onto different commercial rootstocks. Biological Agriculture and Horticulture. 2013;**29**:1-11

[109] Giuffrida F, Cassaniti C, Leonardi C. The influence of rootstock on growth and ion concentrations in pepper (*Capsicum annuum* L.) under saline conditions. The Journal of Horticultural Science and Biotechnology. 2013;**88**:110-116

[110] King SR, Davis AR, Zhang X, Crosby K. Genetics, breeding and selection of rootstocks for solanaceae and cucurbitaceae. Scientia Horticulturae. 2010;**127**:106-111

[111] Mándoki Z, Pénzes B. Effects of using chemical-free root-knot nematode (*Meloidogyne Incognita*) control methods on the occurrence of blossom-end rot in pepper. Journal of Plant Protection Research. 2012;**52**:337-341

[112] Yaoi C, Moreshet S, Aloni B, Karni L. Effect of water stress and climatic factors on the diurnal fluctuation in diameter of bell peppers. The Journal of Horticultural Science and Biotechnology. 2000;**75**:6-11

[113] San Bautista A, Calatayud A, Nebauer SG, Pascual B, Maroto JV, López-Galarza S. Effects of simple and double grafting melon plants on mineral absorption, photosynthesis, biomass and yield. Scientia Horticulturae. 2011;**130**:575-580

[114] Ehret G, Kiemle C, Renger W, Simmet G. Airborne remote sensing of tropospheric water vapor with a near-infrared differential absorption lidar system. Applied Optics. 1993;**32**:4534-4551

[115] Johnson D, Knavel DE. Inheritance of cracking and scarring in pepper fruit. Agriculture. 1990;**115**:172-175

[116] Aloni B, Karni L, Rylski I, Cohen Y. Cuticular cracking in pepper fruit. I. Effect of night temperature and humidity. The Journal of Horticultural Science and Biotechnology. 1998;**73**:743-749

[117] Henry RJ. Genomics strategies for germplasm characterization and the development of climate resilient crops. Frontiers in Plant Science. 2014;**5**:68

[118] Flowers TJ. Improving crop salt tolerance. Journal of Experimental Botany. 2004;**55**: 307-319

[119] Rouphael Y, Schwarz D, Krumbein A, Colla G. Impact of grafting on product quality of fruit vegetables. Scientia Horticulturae. 2010;**127**:172-179

[120] Finckh MR. Sustainable disease management in a European context. In: Collinge DB, Munk L, Cooke BM, editors. Dordrecht: Springer Netherlands; 2008. 399-409 p

[121] Lammerts van Bueren ET, Jones SS, Tamm L, Murphy KM, Myers JR, Leifert C, et al. The need to breed crop varieties suitable for organic farming, using wheat, tomato and broccoli as examples: A review. NJAS—Wageningen Journal of Life Sciences. 2011;**58**(3-4):193-205

[122] Borsani O, Valpuesta V, Botella MA. Developing salt tolerant plants in a new century: A molecular biology. The Plant Cell. 2003;**10**:101-115

[123] Cuartero J, Bolarín MC, Asíns MJ, Moreno V. Increasing salt tolerance in the tomato. Journal of Experimental Botany. 2006;**57**:1045-1058

[124] Martinez-Rodriguez MM, Estañ MT, Moyano E, Garcia-Abellan JO, Flores FB, Campos JF, et al. The effectiveness of grafting to improve salt tolerance in tomato when an "excluder" genotype is used as scion. Environmental and Experimental Botany. 2008;**63**: 392-401

[125] Kim H, Lee K, Hwang H, Bhatnagar N, Kim D-Y, Yoon IS, et al. Overexpression of PYL5 in rice enhances drought tolerance, inhibits growth, and modulates gene expression. Journal of Experimental Botany. 2014;**65**:453-464

[126] Mickelbart MV, Hasegawa PM, Bailey-Serres J. Genetic mechanisms of abiotic stress tolerance that translate to crop yield stability. Nature Reviews. Genetics. 2015;**16**:237-251

[127] Wang W, Vinocur B, Altman A. Plant responses to drought, salinity and extreme temperatures: Towards genetic engineering for stress tolerance. Planta. 2003;**218**:1-14

[128] Munns R. Comparative physiology of salt and water stress. Plant, Cell & Environment. 2002;**25**:239-250

[129] Colla G, Rouphael Y, Leonardi C, Bie Z. Role of grafting in vegetable crops grown under saline conditions. Scientia Horticulturae. 2010;**127**:147-155

[130] Sánchez-Rodríguez E, Romero L, Ruiz JM. Role of grafting in resistance to water stress in tomato plants: Ammonia production and assimilation. Journal of Plant Growth Regulation. 2013;**32**:831-842

[131] Savvas D, Colla G, Rouphael Y, Schwarz D. Amelioration of heavy metal and nutrient stress in fruit vegetables by grafting. Scientia Horticulturae. 2010;**127**:156-161

[132] Penella C, Nebauer SG, Lopéz-Galarza S, SanBautista A, Gorbe E, Calatayud A. Evaluation for salt stress tolerance of pepper genotypes to be used as rootstocks. Journal of Food Agriculture and Environment. 2013;**11**:1101-1107

[133] Pina A, Errea PA. Review of new advances in mechanism of graft compatibility-incompatibility. Scientia Horticulturae. 2005;**106**:1-11

[134] Mudge K, Janick J. A history of grafting. Horticultural Reviews. 2009;**35**:437-494

[135] Lee J-M, Oda M. Grafting of herbaceous vegetable and ornamental crops. In: Horticultural Reviews. 2010. pp. 61-124

[136] Yamakawa K. Use of rootstocks in Solanaceous fruit-vegetable production in Japan. JARQ—Japan Agricultural Research Quarterly. 1982;**15**:175-179

[137] Temperini O, Calabrese N, Temperini A, Rouphael Y, Tesi R, Lenzi A, et al. Grafting artichoke onto cardoon rootstocks: Graft compatibility, yield and Verticillium wilt incidence. Scientia Horticulturae. 2013;**149**:22-27

[138] Trinchera A, Pandozy G, Rinaldi S, Crinò P, Temperini O, Rea E. Graft union formation in artichoke grafting onto wild and cultivated cardoon: An anatomical study. Journal of Plant Physiology. 2013;**170**:1569-1578

[139] Miguel A, De la Torre F, Baixauli C, Maroto JV, Jordá MC, López MM, et al. Injerto de hortalizas. Ministerio de Agricultura, Pesca y Alimentación y Fundación Rural Caja: Valencia, Spain; 2007. 63 p

[140] Lee JM, Kubota C, Tsao SJ, Bie Z, Echevarria PH, Morra L, et al. Current status of vegetable grafting: Diffusion, grafting techniques, automation. Scientia Horticulturae. 2010;**127**:93-105

[141] Santos HS, Goto R. Enxertia em plantas de pimentão no controle da murcha de fitóftora em ambiente protegido. Horticultura Brasileira. 2004;**22**:45-49

[142] Naegele RP, Boyle S, Quesada-Ocampo LM, Hausbeck MK. Genetic diversity, population structure, and resistance to Phytophthora capsici of a worldwide collection of eggplant germplasm. PLoS One. 2014;**9**:e95930

[143] Penella C, Nebauer SG, López-Galarza S, San Bautista A, Rodríguez-Burruezo A, Calatayud A. Evaluation of some pepper genotypes as rootstocks in water stress conditions. Horticultural Science. 2014;**41**:192-200

[144] López-Serrano L, Penella C, San-Bautista A, López-Galarza S, Calatayud A. Physiological changes of pepper accessions in response to salinity and water stress. Spanish Journal of Agricultural Research. 2017;**15**:e0804

[145] Zhu J, Bie Z, Huang Y, Han X. Effect of grafting on the growth and ion concentrations of cucumber seedlings under NaCl stress. Soil Science & Plant Nutrition. 2014;**54**:895-902

[146] Aloni B, Cohen R, Karni L, Aktas H, Edelstein M. Hormonal signaling in rootstock-scion interactions. Scientia Horticulturae. 2010;**127**:119-126

[147] Ferreira-Silva SL, Silva EN, Carvalho FEL, de Lima CS, Alves FAL, Silveira JAG. Physiological alterations modulated by rootstock and scion combination in cashew under salinity. Scientia Horticulturae. 2010;**127**:39-45

[148] Penella C, Nebauer SG, Quiñones A, San Bautista A, López-Galarza S, Calatayud A. Some rootstocks improve pepper tolerance to mild salinity through ionic regulation. Plant Science. 2015;**230**:15-12

[149] Penella C, Nebauer SG, López-Galarza S, Quiñones A, San Bautista A, Calatayud Á. Grafting pepper onto tolerant rootstocks: An environmental-friendly technique overcome water and salt stress. Scientia Horticulturae. 2017;**226**:33-41

[150] Guidi L, Landi M, Penella C, Calatayud A. Application of modulated chlorophyll fluorescence and modulated chlorophyll fluorescence imaging to study the environmental stresses effect. Annali di Botanica. 2016;**6**:39-56

[151] Rouphael Y, Cardarelli M, Rea E, Colla G. Improving melon and cucumber photosynthetic activity, mineral composition, and growth performance under salinity stress by grafting onto Cucurbita hybrid rootstocks. Photosynthetica. 2012;**50**:180-188

[152] Nilsen ET, Freeman J, Grene R, Tokuhisa J. A Rootstock provides water conservation for a grafted commercial tomato (*Solanum lycopersicum* L.) line in response to mild-drought conditions: A focus on vegetative growth and photosynthetic parameters. PLoS One. 2014;**9**:e115380

[153] Kumar P, Rouphael Y, Cardarelli M, Colla G. Vegetable grafting as a tool to improve drought resistance and water use efficiency. Frontiers in Plant Science. 2017;**8**:1130

[154] Martínez-Ballesta MC, Alcaraz-López C, Muries B, Mota-Cadenas C, Carvajal M. Physiological aspects of rootstock–scion interactions. Scientia Horticulturae. 2010;**127**: 112-118

[155] Ruiz JM, Ríos JJ, Rosales MA, Rivero RM, Romero L. Grafting between tobacco plants to enhance salinity tolerance. Journal of Plant Physiology. 2006;**163**:1229-1237

[156] Liu J, Li J, Su X, Xia Z. Grafting improves drought tolerance by regulating antioxidant enzyme activities and stress-responsive gene expression in tobacco. Environmental and Experimental Botany. 2014;**107**:173-179

[157] Liu S, Li H, Lv X, Ahammed GJ, Xia X, Zhou J, Shi K, Asami T, Yu J, Zhou Y. Grafting cucumber onto luffa improves drought tolerance by increasing ABA biosynthesis and sensitivity. Scientific Reports. 2016;**6**:20212

[158] López-Marín J, Gálvez A, Del Amor FM, Albacete A, Fernández JA, Egea-Gilabert C, et al. Selecting vegetative/generative/dwarfing rootstocks for improving fruit yield and quality in water stressed sweet peppers. Scientia Horticulturae. 2017;**214**:9-17

[159] Penella C, Nebauer SG, Bautista AS, López-Galarza S, Calatayud T. Rootstock alleviates PEG-induced water stress in grafted pepper seedlings: Physiological responses. Journal of Plant Physiology. 2014;**171**:842-851

Chapter 8

Bringing Climate Smart Agriculture to Scale: Experiences from the Water Productivity Project in East and Central Africa

Kizito Kwena, Fitih Ademe, Joseph Serge,
Nezeghty Asmerom, Bernard Musana,
Razaka Razakamiaramanana, Reuben Ruttoh,
Hezron Mogaka, Assefa Dereje, Kifle Woldearegey,
Anthony Esilaba and Rosemary Emongor

Additional information is available at the end of the chapter

http://dx.doi.org/10.5772/intechopen.72365

Abstract

Since 2010, six research organizations in the region have implemented a regional project that sought to combat food insecurity, poverty and climate change by up-scaling Climate-Smart Agriculture (CSA) technologies across farms and landscapes using the Climate Smart Landscape (CSL) approach. Several CSA technologies were evaluated and promoted across landscapes using this approach with remarkable success. Maize yields in Kenya rose from 0.5 to 3.2 t ha^{-1}, resulting in over 90% of the watershed communities being food secure. In Madagascar, rice yields increased from 2 to 4 t ha^{-1} whilst onion yields increased from 10 to 25 t ha^{-1}, resulting in watershed communities being 60% food-secure. In Eritrea, sorghum yields increased from 0.6 to 2 t ha^{-1}. Farmers in Ethiopia earned US$10,749 from the sale of pasture whilst in Madagascar, watershed communities earned additional income of about US$2500/ha/year from the sale of onions and potatoes during off-season. Adoption levels of various CSA technologies rose from less than 30% to over 100% across the participating countries, resulting in rehabilitation of huge tracts of degraded land. In a nutshell, the potential for CSL in the region is huge and if exploited could significantly improve our economies, lives and environment.

Keywords: climate change, Climate-Smart Landscapes, Climate-Smart Agriculture, innovation platform, food security

1. Introduction

The East and Central Africa (ECA) subregion is projected to getting warmer and wetter by the end of this century. Temperatures are projected to increase by about 2°C and rainfall by about 11% by 2050 [1, 2]. It is therefore possible that the subregion could be food self-sufficient because of climate change. As unfamiliar as this counter-narrative might seem, climate change presents an opportunity for the subregion to think and act differently, to change the way it views growth and interacts with the environment, and to choose a different path toward sustainable development. The Zero Hunger by 2025 target set by African Heads of State is achievable. However, this will only be possible if countries in the subregion invest 10% of their GDP in agriculture and target to grow the sector by 6% as proposed by the African Unions' Comprehensive Africa Agricultural Development Programme (CAADP) in 2003. So how does the subregion get there? By making substantial investments in Climate-Smart Agriculture (CSA). Climate-Smart Agriculture, if adopted, has the potential to usher in a new era of clean and sustainable growth for the subregion.

Climate-Smart Agriculture is an applied set of farming principles and practices that increases productivity in an environmentally and socially sustainable way (adaptation), strengthens farmers' capacities to cope with the effects and impacts of climate change (resilience), conserves the natural resource base through maintaining and recycling organic matter in soils (carbon storage), and as a result reduces greenhouse gas emissions (mitigation) [3]. This approach also aims to strengthen livelihoods and food security, especially of smallholders, by improving the management and use of natural resources and adopting appropriate methods and technologies for the production, processing and marketing of agricultural goods [1, 3–5]. However, for agricultural systems in the subregion to achieve CSA objectives, including improved food security and rural livelihoods as well as climate change adaptation and mitigation, they need to take a landscape approach; they must become 'Climate-Smart Landscapes.' Climate-Smart Landscapes (CSL) operate on the principles of integrated watershed management (IWM) while explicitly incorporating adaptation and mitigation into their management objectives [1, 4].

For 3 years, the Kenya Agricultural and Livestock Research Organization (KALRO), Rwanda Agricultural Board (RAB), Eritrea's National Agricultural Research Institute (NARI), Ethiopian Institute of Agricultural Research (EIAR), Artelia Madagascar (AMG) and Madagascar's Centre National de Recherché Applique au Developpement Rural (FOFIFA) implemented a regional project on improving agricultural water productivity using this approach. The project sought to combat food insecurity, poverty and climate change by increasing the availability and productivity of water in smallholder rain-fed and irrigated agriculture at both farm and landscape levels.

The project was implemented from 2010 to 2013 in five countries namely Kenya, Rwanda, Eritrea, Madagascar and Ethiopia with financial support from the Association for Strengthening Agricultural Research in East and Central Africa (ASARECA) and her partners. Due to positive results from this project, a second phase was launched in 2014 and implemented up to 2015 in three more countries (Uganda, Sudan and Burundi) with the aim of up-scaling 'best bet' CSA technologies from the first phase and establishing more CSL. This chapter seeks to highlight some of the benefits of CSL and its potential in the region with a view to encouraging governments to invest in this noble approach to agricultural development in order to combat food insecurity, poverty and climate change.

2. Methodology

To establish and successfully promote and sustain climate-smart agricultural landscapes that could generate important synergies for agricultural production, climate adaptation and mitigation, as well as other livelihood and environmental objectives at farm and landscape scales, the following activities were undertaken.

2.1. Selecting the watersheds/landscapes

Two watersheds measuring about 100 km^2 were identified in each country by all stakeholders during the national stakeholders' consultative workshops conducted prior to project inception. The two watersheds were selected based on the extent of their degradation, potential to benefit from improved water management, their vulnerability to climate variability and change, and their food security and poverty levels. Mwania and Kalii watersheds in Machakos and Makueni counties, respectively, were selected for Kenya; Karama and Muse-Bivumu in Nyamagabe and Bugesera districts, respectively, in Rwanda; Adulala and Ketchema in Ethiopia; Amadir and Molqi in Eritrea; and Ankazomiriotra and Avaratrambolo in Mandoto and Manjakandriana districts, respectively, in Madagascar (**Figure 1**).

These watersheds were all densely populated, highly degraded, food insecure and very prone to high climatic stresses. They therefore presented huge opportunity for CSL to improve agricultural production, resilience and income of their communities through the use of appropriate and available CSA technologies. The sites also had many agricultural development initiatives which complemented CSL efforts. They also had a lot of secondary data on climate, land and water resources, crop production and demographic trends which facilitated long-term planning and accurate simulation of climate change impacts. Finally, they had good land tenure systems which allowed farmers to invest in long-term and capital-intensive CSA practices such as drip irrigation, agroforestry, CA, terracing and water pans across the landscapes.

2.2. Conducting the baseline survey

A comprehensive baseline study was conducted at the start of the project to capture the socioeconomic situation, resource availability, average production and income, adaptation, mitigation, biodiversity conservation and risk management approaches of village households before the project. This was done to generate indicators for monitoring the impact of CSA interventions up-scaled across the landscapes by the project and to encourage investment in CSL.

Figure 1. Location of climate-smart landscapes in Kenya, Eritrea, Ethiopia, Rwanda and Madagascar.

2.3. Forming multi-stakeholder platforms

The project established Innovation platforms in each watershed in which all stakeholders with interest in the watershed were brought together and made part of the project implementation team. They were briefed on the objectives of CSL to secure their buy-in. This was done to consolidate resources, share knowledge, build coalitions and pool investments. The stakeholders were drawn from the watershed communities, local administration, non-governmental organizations (NGOs), government departments, religious groups, donor agencies, agrodealers and financial institutions. They were all involved in landscape planning, project implementation and progress monitoring for CSL objectives, as well as others. Landscape management plans with clearly defined roles and responsibilities were developed to guide this process.

2.4. Prioritizing and up-scaling CSA interventions

As indicated before, the project adopted the CSL approach to resolve the problem of land degradation, food insecurity and poverty in the six watersheds. Climate-Smart Landscapes, like the IWM approach, link production, conservation and livelihood objectives of people with a stake in a given landscape/watershed. It provides a framework for integrating technical, economic and social knowledge in identifying constraints and in supporting planning and decision-making to achieve sustainable solutions. Through this approach, numerous CSA technologies were evaluated and promoted across landscapes using field demonstrations, field days, farmer exchange visits and trainings. The technologies were selected by farmers based on their ease of adoption, investment required and ability to make best use of increased water availability. These included conservation agriculture (CA), agroforestry, manure management, water harvesting, terracing, mulching, drought-tolerant crops, proper agronomy, high-productivity crop varieties and use of weather-based agroadvisories.

2.5. Building capacity of stakeholders

Capacity of communities was strengthened to enhance adoption and utilization of CSA technologies. The project held numerous meetings to sensitize stakeholders on the benefits of CSA and CSL. Field experiments were also conducted to demonstrate the complete portfolio of CSA interventions and to generate more scientific evidence to support CSA. The project, private sector and local governments also organized regular training sessions for farmers on good agricultural practices.

2.6. Monitoring and evaluation

To attract more interest and investment in CSL, the project developed a comprehensive monitoring framework which captured the multiple benefits of CSL which included yield improvements, food and energy security, adaptation, mitigation, human health, biodiversity conservation and other ecosystems services. Farmers also maintained a daily diary of their farm activities and worked with the project staff to monitor and evaluate the progress of their chosen interventions. These results were digitized and analyzed by researchers and discussed by all stakeholders at the end of every crop season.

2.7. Dissemination of outcomes

Participatory videos on success stories and testimonials from the pilot landscapes were screened in nearby watersheds to spread the message of CSL. Success stories were also widely publicized through local, national and international media. The project also organized regular farmer field days and exchange visits to motivate farmers, address their questions and improve on existing strategies.

3. Results

3.1. Food security

Food security is a major challenge for the East and Central Africa (ECA) subregion. ECA is among the few regions in the world where yields have been stagnant over the past 50 years, leading to a decline in per capita food production and malnutrition. From the baseline surveys conducted at project inception, many households in all the five countries experienced serious food insecurity for many months in a year. In Kenya, for instance, over 50%

of the household in both watersheds lacked sufficient food to feed their families and relied on food aid. The situation was the same in Ethiopia, Madagascar and Eritrea where over 44, 45 and 55%, respectively, of the households were food insecure. However, through project intervention, productivity in all the watersheds/landscapes increased significantly and most watershed communities are now food secure. In Kenya, for instance, by embracing forecast-based farming, tied ridging, seed priming, improved agronomic practices, improved crop varieties, and micro-dosing among other technologies, farmers posted good yields throughout the project period despite most seasons being bad. Maize yields ranged from 1.2 to 3.2 t ha^{-1} compared to baseline yield of less than 0.5 t ha^{-1} (**Figure 2**). Hence, most households (hh) in the two watersheds, 3600 hh or over 90%, are food secure.

In Madagascar, adoption of improved rice varieties increased rice yields from 2 to 4 t ha^{-1} while onion yields increased from 10 to 25 t ha^{-1} due to prudent management of water and other inputs. As a result, communities in Ankazomiriotra and Avaratrambolo watersheds are now 60% food secure. In Eritrea, sorghum yields increased from 0.6 t ha^{-1} at project inception to 1.5–2 t ha^{-1} due to soil and water conservation (SWC) initiatives.

3.2. Increased income

A dominant feature of the ECA is widespread poverty and malnutrition. Majority of the people in the subregion, including all the watersheds, live in abject poverty. In Machakos and Makueni counties in Kenya, for instance, about 52 and 64% of the population, respectively, live below the poverty line (on less than US$ 1 per person per day). However, through CLS approach, the situation in all the watersheds improved markedly. In Ethiopia, for instance, farmers in Adulala were able to harvest 102 kg of honey worth about US$ 568 in one season from 10 out of 28 beehives set up by the project. About 22 households benefitted from these proceeds, and the income is bound to increase with time as more hives get colonized. Farmers in Adulala also managed to harvest and sell pasture/grass worth US$ 10,749 from the hillside rehabilitation activity. A total of 720 farmers benefitted from these proceeds.

In Madagascar, watershed communities are now able to earn additional income of about US$ 2500 ha^{-1} yr^{-1} from the sale of onions and potatoes during off-season due to prudent

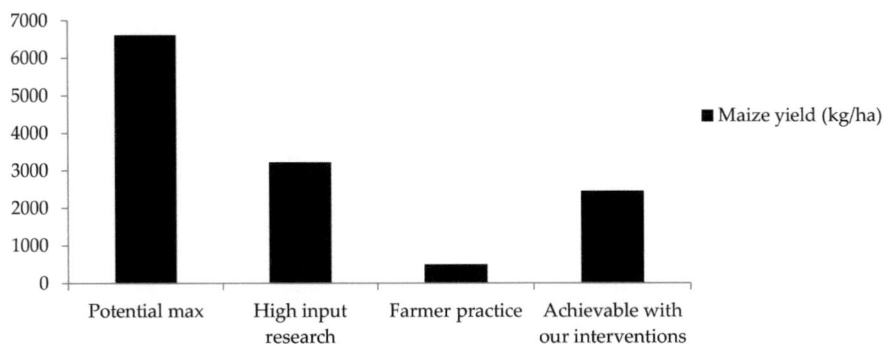

Figure 2. Effect of CSA interventions on maize yields in Kenya.

management of water and other inputs. Similarly, in Eritrea, each of the 66 out of 480 households who adopted agroforestry was able to earn about US$ 450 in just 6 months from the sale of *Rhamnus* leaves and vegetables. Most of them used this money to buy sheep and poultry to diversify and increase income.

3.3. Ecosystem improvement

Low adoption of productivity-enhancing technologies has widely been blamed for low agricultural productivity in sub-Saharan African. From the baseline surveys conducted in the five countries, most farmers were knowledgeable about CSA practices but did not adopt and use them. In Mwania watershed in Kenya, for instance, 77 and 87% of farmers were knowledgeable about irrigation and tied ridges but only 18 and 16% practiced the technologies, respectively. However, awareness and use of terraces were the highest in both sites with 98.9 and 87.1% in Mwania and Makindu, respectively (**Table 1**).

The low level of adoption of terraces in Makindu was due to the relatively flat landscape compared to Mwania. Landscape at Mwania is hilly, and slopes often exceed 25%, making it essential to use structures such as terraces. Similarly, high level of adoption of irrigation and tied ridges in Makindu compared to Mwania was due to availability of water and ease with which it could be applied. Tied ridging is labor intensive, and this could be the reason behind low usage of this technology in both Mwania and Makindu locations. Various models have been used to deliver these technologies to farmers with very minimal success. However, through the CSL approach adopted by this study, several CSA technologies were up-scaled with very positive results. In Kenya, for instance, out of 198 farmers trained on terracing to conserve soil and water and improve productivity, over 700 constructed them on their farms and realized very good maize yields. Similarly, of the 146 farmers trained on pitting to harvest runoff and grow fodder, over 600 managed to dig over 50,000 pits on their farms and plant Napier grass for their livestock. The extra adopters learnt from their neighbors who attended the trainings. As a result, huge tracts of degraded land have been rehabilitated and over 100 tonnes of pasture produced compared to zero at inception. Farmers have been able to sell them and earn extra income.

Technology	Knowledge (%)**	Usage (%)**	Mwania		Makindu	
			Knowledge (%)	Usage (%)	Knowledge (%)	Usage (%)
Conservation farming	72.4	44.8	68.9	31.2	75.1	58.3
Irrigation	81.9	30.6	76.8	18	86	43.2
Mulching	66.1	32.2	75.2	27.3	58.9	37
Terraces	98.9	87.1	98.8	98.1	99	76.1
Tied ridges	74.1	41.1	87	16	74.1	66.2
Water harvesting	85.3	53.8	91.6	62.4	80.1	45.1

**Significance at $p \leq 0.01$.

Table 1. Technology knowhow and use.

Benefits	Mwania	Makindu	Mean
Decreased runoff and erosion (%)	78	83	81
Increased water infiltration (%)	26	86	56
Improved soil moisture conditions (%)	43	53	48
Improved soil physical properties (%)	—	75	38

Table 2. Benefits of investments in SWC technologies.

Majority of the technologies adopted were mainly for soil and water conservation (SWC) and were preferred because of their perceived benefits. The benefits included decreased runoff and erosion (81%), increased water infiltration (56%), improved soil moisture conditions (48%) and improved soil physical properties (38%) as shown in **Table 2**. A study conducted in the two landscapes/watersheds to compare the rate of adoption of these and other CSA technologies between male- and female-managed farms established that there was no difference in the adoption rate between male- and female-managed farms in the two watersheds; however, male-managed farms preferred capital-intensive technologies such as irrigation while female-managed farms adopted labor and capital-reductive technologies such as conservation agriculture [6].

Finally, farmers in Rwanda and Kenya established nurseries and planted over 1.5 million tree seedlings on their farms to improve the environment and generate income.

4. Conclusions

The high level of adoption of CSA practices in landscapes/watersheds across the countries clearly indicates that available CSA technologies are acceptable to farmers if the same are tailored to meet the needs and requirements of the farmers with due consideration to their biophysical and socioeconomic conditions compared to generalized recommendations targeting a given agroecology or administrative unit. Another important finding of this work is that mobilizing communities and enhancing their capacity to better understand the tangible and intangible benefits from CSL and CSA interventions has much bigger impact than dealing with individual farmers. The landscape/watershed committees and innovation platforms established under this project played a vital role in increased adoption of CSL in all target countries.

In a nutshell, the potential for CSL approach and its benefits in the region are huge. However, to successfully transit from CSA to CSL: (1) all stakeholders in a given watershed/landscape must be involved in the planning, implementation and monitoring of this transition; (2) a comprehensive monitoring framework that clearly indicates the socioeconomic and environmental benefits of CSL must be developed, and the results communicated to stakeholders regularly to attract more investment in CSL; (3) the sites must have many ongoing agricultural development initiatives to complement and reduce the cost of establishing CSL; a lot of secondary data on climate, land and water resources, crop production and demographic trends to facilitate long-term planning and accurate simulation of climate change impacts; and good land tenure systems to enable farmers invest in long-term and capital-intensive CSA practices; (4) massive civic education and capacity building are required to educate stakeholders on the benefits of CSL; (5) ready market must be available to absorb increased agricultural yields from CSL; and (6) landscape communities must embrace weather-based agroadvisories to minimize risks posed by climate variability and promote investment in CSL.

Acknowledgements

The authors are grateful to ASARECA and her partners for the financial support. We are also grateful to the landscape/watershed communities and other stakeholders in the five countries for participating in this study.

Author details

Kizito Kwena[1*], Fitih Ademe[2], Joseph Serge[3], Nezeghty Asmerom[4], Bernard Musana[5], Razaka Razakamiaramanana[6], Reuben Ruttoh[1], Hezron Mogaka[7], Assefa Dereje[8], Kifle Woldearegey[8], Anthony Esilaba[9] and Rosemary Emongor[10]

*Address all correspondence to: kwenakizito@yahoo.com

1 Kenya Agricultural and Livestock Research Organization (KALRO), Katumani Research Centre, Machakos, Kenya

2 Ethiopian Institute of Agricultural Research (EIAR), Melkasa Agricultural Research Centre, Nazareth, Ethiopia

3 Artelia Madagascar, Antananarivo, Madagascar

4 National Agricultural Research Institute (NARI), Asmara, Eritrea

5 Rwanda Agricultural Board (RAB), Kigali, Rwanda

6 Centre National de Recherché Applique au Developpement Rural (FOFIFA), Antsirabe, Madagascar

7 Environmental Research & Policy Analysis Group, Nairobi, Kenya

8 Mekelle University, Mekelle, Ethiopia

9 Kenya Agricultural & Livestock Research Organization (KALRO), Nairobi, Kenya

10 Kenya Agricultural & Livestock Research Organization (KALRO), National Agricultural Research Laboratories, Nairobi, Kenya

References

[1] Food and Agriculture Organization of the United Nations (FAO). Climate-Smart Agriculture Sourcebook. 1st ed. Rome, Italy: FAO; 2013. 570 p. ISBN: 978-92-5-107720-7, E-ISBN978-92-5-107721-4

[2] World Bank, editor. Climate-Smart Agriculture: A call to Action. 1st ed. Washington DC, USA: World Bank Group; 2011. 28 p. http://documents.worldbank.org/curated/en/992021468197391264/Climate-smart-agriculture-a-call-to-action

[3] Food and Agriculture Organization of the United Nations (FAO). Climate-Smart Agriculture: Policies, practices and financing for food security, adaptation and mitigation. In: FAO, editor. Conference on Agriculture, Food Security and Climate Change; 31 October 2010-5 November 2010; Hague, Netherlands. Rome, Italy: FAO; 2010. p. 48. www.fao.org/docrep/018/i3325e.pdf

[4] Scherr SJ, Shames S, Friedman S. From climate-smart agriculture to climate-smart landscapes. Agriculture and Food Security. 2012;**1**(12). DOI: 1186/2048-7010-1-12

[5] Lipper L, Thornton P, Campbell BM, Baedeker T, Braimoh A, Bwalya M, Caron P, Cattaneo A, Garrrity D, Henry K, Hottle R, Jackson L, Jarvis A, Kassam F, Mann W, McCarthy N, Meybeck A, Neufeldt H, Remmington T, Thi Sen P, Sessa R, Shula R, Tibu A, Torquebiau E. Climate-Smart Agriculture for food security. Nature Climate Change. 2014;**14**:1068-1072. DOI: 10.1038/nclimate2437

[6] Jay K. Gender Disaggregated Analysis of Adoption of Agricultural Water Management Technologies in Eastern Africa [Thesis]. London, UK: University College of London; 2012. 76 p

Chapter 9

Climate Control in Mediterranean Greenhouses

A. Nafi Baytorun and Zeynep Zaimoglu

Additional information is available at the end of the chapter

http://dx.doi.org/10.5772/intechopen.72367

Abstract

As climate control in greenhouses directly affects crop yields, there is an increasing trend for advancements in environmentally controlled agricultural-production techniques. In the Mediterranean region, the temperatures during the period from December to February are below 12°C when the daily total radiation 8.4 MJ/m^{2}day. Based on the region's climate data, greenhouses require heating during the period from November to March, ventilation and shading from February to May and cooling from June to September. In order to maintain day and night temperatures of 18/16°C, annual heat energy requirement of PE greenhouses is 95-256 kWh/m^2. In view of environment and production costs, conservation of heating energy is as important as heating itself. Heat energy saving is about 37% when energy curtains are used. Greenhouse temperature can be increased by 8°C in palliative non-heated greenhouses where energy curtains and water mattresses are used in addition to passively used solar energy. Ventilation openings at the roofs of these greenhouses should adequately be 20-25%. When outside noon-time temperature is above 30°C in June, evaporative cooling of greenhouse is essential. Depending on outside humidity and volume of exchanged air for cooling, a temperature difference of 6°C can be achieved with evaporative cooling of greenhouses in August.

Keywords: greenhouse heating, energy saving, ventilation, cooling

1. Introduction

Countries need to increase the efficiency and quality of their agricultural production in order to meet their future requirements in line with population increase. In our country, it has become necessary to take particular measures due to rapid population increase and globalization of trade. Growing fruits and vegetables in controlled environments with low production costs is among these measures to be taken.

Controlling environmental conditions in agricultural production has direct influence on efficiency. For this reason, environmentally controlled agricultural techniques have developed at an increasing rate. In environmentally controlled plant production systems, it is aimed to change natural environmental factors according to the optimum requirements of plants. The most common and effective implementation of environmentally controlled plant production takes place in greenhouses.

The objective of innovative technologies in greenhouses is to improve "Quality of Life Cycles." Therefore, in order to achieve sustainability, it is highly important to correctly analyze the inputs and outputs required for the efficiency to be obtained from a unit area in greenhouses [1].

Sustainable greenhouse systems should be equipped with resource conserving, socially supported, commercial, competitive, environmentally friendly, reliable production technologies, and they should aim to reduce energy, water and chemical pesticide requirements besides avoiding waste production [2].

Greenhouse practices in Turkey started first in the 1940s in the Mediterranean region, particularly in Antalya, and then spread to the Aegean and Marmara regions depending on ecological conditions. The Mediterranean region has 84% of the country's total greenhouse area, followed by the Aegean, Black Sea and Marmara regions with 9.4, 4.8 and 1.7%, respectively.

With 22,000 ha, Antalya has 37% of the country's total greenhouse area. A total of 32,000 ha of the 61,500 ha greenhouse area in the country consists of greenhouses, which are defined as high greenhouse systems [3, 4]. As big investment groups entered the sector, modern greenhousing has rapidly developed and reached a level of 1000 ha. This figure increases by 150–200 ha every year. Nowadays, modern greenhousing is practiced in 3% of our total greenhouse area, and in the following decade, this share is expected to reach 15% [5].

In terms of equipment and technology, greenhouses in the Mediterranean region can be divided into two groups:

1. *Greenhouses with low technology:* These greenhouses have simple iron structures. They are covered with PE plastic or mixed PE plastic. Starting from 1987, galvanized pipes have been used in greenhouses built with government support provided within the framework of Resource Utilization Support Fund (RUSF). These greenhouses with ventilation openings only on the sidewalls have been built as blocks consisting of four sections. In all greenhouses installed with RUSF support, drop irrigation systems are used, and these greenhouses are heated on a regular basis.

2. *Modern greenhouses with high technology:* These greenhouses require quite high initial investment costs, which are considered as big businesses. They are generally built with galvanized steel and aluminum materials. As cover material glass, mixed PE plastic or double pane PC is used. They are projected as high volume structures. In these types of greenhouses, modern production techniques like soilless agriculture are used. Their irrigation systems are projected as computer-aided closed systems, and they have central heating systems. In addition to natural ventilation, these greenhouses have fans that provide air circulation. Greenhouses, in which ornamental plants are grown, have evaporative cooling systems. During hot periods, besides cooling, moving shading systems are activated.

2. Climatic values in the Mediterranean region

The most important climatic parameters in greenhouse cultivation are solar radiation, day length and temperature values. Solar radiation is one of the most important climatic parameters to be taken into consideration in a location where a greenhouse is going to be built. Total daily solar radiation in a greenhouse location should be minimum 8.5 MJ/m^2 day (2.34 kWh/m^{2}day) [6]. However, insolation time is as important as solar radiation. Products grown in greenhouses require an average of 6 h of day length. In other words, during months with short day length (e.g., November, December and January), total day length should be minimum of 500–550 h [6].

Total annual insolation time and intensity of radiation in Turkey are 2.640 h (7.2 h/day) and 1.311 kWh/m^{2}year (3.6 kWh/m^{2}day), respectively. In **Tables 1** and **2**, long-year average total daily radiation values and insolation time for some cities on the Mediterranean coastline are given. As can be seen from the tables, insolation time and total daily solar radiation values in the Mediterranean region are above average values in whole Turkey.

Vegetables grown in greenhouses have adapted to an average temperature of 17–28°C. Products grown in greenhouses undergo stress at temperatures below 12°C and above 32°C. At low temperature values like frost, irreversible harms occur. Greenhouses should be heated when the outside temperature falls below 12°C. When outside temperature is between 7 and 12°C, heating is necessary only during night hours. For desirable plant growth, the difference between night and day temperatures should be 5–7°C [6].

City	January	February	March	April	May	June	July	August	September	October	November	December
Adana	1.98	2.42	4.12	4.98	6.07	6.68	6.46	5.91	4.90	3.78	2.33	1.81
Antalya	2.12	2.57	4.37	5.47	6.36	6.93	6.65	6.14	5.16	3.93	2.51	1.92
Hatay	1.99	2.42	4.01	4.87	5.96	6.63	6.31	5.82	4.75	3.63	2.35	1.79
İçel	2.11	2.65	4.27	5.24	6.28	6.86	6.66	6.08	5.04	3.84	2.47	1.91
Muğla	2.11	2.42	4.24	5.40	6.22	6.81	6.47	6.05	5.05	3.96	2.56	1.88
Turkey	1.79	2.50	3.87	4.93	6.14	6.57	6.50	5.81	4.81	3.46	2.14	1.59

Table 1. Total daily radiation values in different cities in the Mediterranean region (kWh/m^{2}day).

City	January	February	March	April	May	June	July	August	September	October	November	December
Adana	4.67	5.65	6.97	7.84	9.72	11.29	11.17	11.22	10.15	7.78	5.86	4.21
Antalya	4.95	6.10	7.24	8.29	9.70	11.55	11.84	11.29	9.80	7.68	5.97	4.55
Hatay	5.09	6.22	7.17	8.28	10.23	11.14	10.89	10.47	9.80	7.86	6.37	4.99
İçel	4.99	6.04	7.35	8.38	9.94	11.18	11.45	11.03	10.02	7.91	6.15	4.64
Muğla	5.13	6.20	7.12	8.18	9.91	11.73	11.90	11.31	9.92	7.85	6.01	4.67
Turkey	4.11	5.22	6.27	7.46	9.10	10.81	11.31	10.70	9.23	6.87	5.15	3.75

Table 2. Insolation time in different cities in the Mediterranean region (h).

At temperatures between 12 and 22°C, by using passive acclimatization (natural ventilation), it is possible to arrange the greenhouse environment according to the values required by plants. When outside temperature exceeds 27°C, it is necessary to install highly expensive cooling systems in greenhouses. The greenhouses on the Mediterranean region should be left idle during the specified periods.

The graphical representation of long year average temperatures and total daily radiation values of Mediterranean cities Antalya (36°:53′ N), Mersin (36°:48′ N) and Hatay (36°:14′ N) is given in **Figure 1**. As seen in the figure, total daily radiation in all these three cities on the Mediterranean coastline is below 2.3 kWh/m^{2}day (8.4 MJ/m^2 day) during December and January. This indicates that solar radiation is insufficient for plant growth during these 2 months. In order to allow more solar radiation into the greenhouse in December and January, greenhouse roofs should be covered using a material with high impermeability. The 1% decrease in intensity of light reaching the greenhouse results in the same decrease in efficiency.

Another factor that affects plant growth is day length. Total day length in Antalya during November–January in Antalya is 474 h. This value is close to the limit value, which is accepted as 500–550 h.

An overview of the temperature values of cities in the Mediterranean region (**Table 3**) shows that average daily temperature during the period between December and February is below 12°C. However, as average temperature during these months does not fall below 7°C, producers in this region prefer cold greenhousing and take simple heating measures in order to continue production during very cold days.

One of the main problems in unheated PE plastic greenhouses in the Mediterranean region is that greenhouse temperature falls below outside temperature on nights when the sky is clear. This results from the fact that PE plastic transmits long-wave heat rays on a specific band. In their measurements taken in glass and PE plastic greenhouses in the Mediterranean region

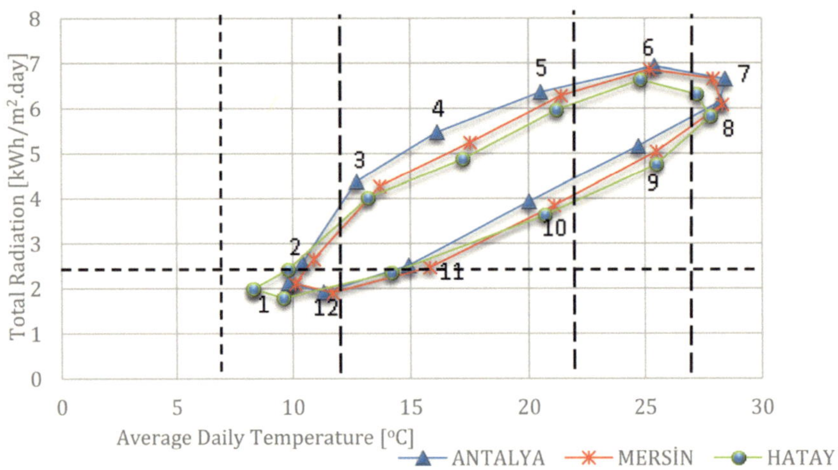

Figure 1. Average daily temperatures and total radiation values in three cities in the Mediterranean region: Antalya (36°:53′ N), Mersin (36°:48′ N) and Hatay (36°:14′ N).

City	January	February	March	April	May	June	July	August	September	October	November	December
Adana	9.6	10.6	13.6	17.6	21.7	25.6	28.2	28.6	26.1	21.8	15.7	*11.1*
Hatay	8.1	9.8	13.2	17.1	21.1	24.8	27.2	27.8	25.7	20.8	14.2	*9.5*
Antalya	9.9	10.4	12.8	16.1	20.3	25.1	28.2	28.0	24.9	20.3	15.1	*11.4*
Mersin	10.3	11.1	13.9	17.4	21.1	24.8	27.7	28.2	25.7	21.5	16.1	*11.9*
Muğla	5.3	5.9	*8.6*	12.5	17.7	22.9	26.3	26.1	21.8	16.1	*10.5*	*6.8*

Table 3. Average daily temperature values of different cities in the Mediterranean region (°C).

Cover material	Outside temperature (°C)	Inside temperature (°C)	$(t_i - t_o)$ (°C)
Plastic	9.3	9.0	−0.3
	10.3	9.0	−1.3
	8.1	8.1	0.0
Glass	7.3	9.1	1.8
	6.8	7.7	0.9
	8.3	10.1	1.8

Table 4. Inside and outside temperature values recorded on different days in an unheated PE plastic greenhouse (time 17:00–07:00 Average values).

(Adana), it is observed that the temperature falls below outside temperature in PE greenhouses without thermal curtains [7]. This situation was not seen in unheated glass greenhouses (**Table 4**).

In order to achieve year-long production in greenhouses in the Mediterranean region, greenhouses should be heated during night hours in the winter, ventilated and shaded in transition periods and cooled during hot periods. As cooling is a very costly acclimatization measure, it is not a preferable method for greenhouses, which are usually left idle during such periods.

2.1. Heating

Total daily solar energy values for Antalya (36°:53′ N) in different months and required daily heat energy values for PE plastic and glass greenhouses for specific temperatures are given in **Table 5**. As can be seen from **Table 6**, total daily solar energy reaching the greenhouse in Mediterranean climatic condition exceedingly meets the heat energy requirement in all months of the year. However, only some of the solar energy reaching the greenhouse during day hours can be stored. Heat losses that occur through the cover material immediately after sunset lead to rapid decreases in greenhouse temperature.

Heating in greenhouses has a significant effect on production costs. Heat energy requirements of plastic greenhouses installed in the Mediterranean region vary depending on climate of the location, greenhouse type and greenhouse equipment.

Months	January	February	March	April	May	June	July	August	September	October	November	December
Q_{sun}	2.12	2.57	4.37	5.47	3.37	6.93	6.65	6.0	5.17	3.93	2.52	1.92
							PE plastic greenhouse					
$Q_{(16)}$	1.06	0.94	0.57	0.19	0.00	0.00	0.00	0.00	0.00	0.00	0.34	0.79
$Q_{(17/18)}$	1.27	1.13	0.73	0.31	0.00	0.00	0.00	0.00	0.00	0.04	0.49	0.98
							Glass greenhouse					
$Q_{(16)}$	1.04	0.92	0.53	0.14	0.00	0.00	0.00	0.00	0.00	0.00	0.28	0.75
$Q_{(17/18)}$	1.27	1.13	0.69	0.26	0.00	0.00	0.00	0.00	0.00	0.00	0.44	0.96

Table 5. Daily solar energy reaching the greenhouse in Antalya climatic conditions required daily heat energy values for PE plastic and glass greenhouses for different temperatures (kWh/m^{2}day).

Greenhouse requirement	Heat energy requirement kWh/m^2 a				
	Antalya	Adana	Mersin	Hatay	Muğla
Single pane plastic	126.6	113.6	95.5	140.2	256.3
Roof single, side wall double pane PE plastic	118.3	106.3	89.3	130.6	239.2
Roof single, side wall double pane PE plastic + Thermal curtain moderately insulated	95.1	87.4	72.7	107.7	198.3

Table 6. Heat energy requirements during the production year for different PE plastic greenhouses in some cities in the Mediterranean region when night/day temperature is kept at 16/18°C [8].

Daily heat energy values based on climatic conditions of Antalya for weeks of the year when greenhouse temperature is kept at 18/16°C, 18/14°C and 16/12°C are given in **Figure 2**. As can be seen from the figure, daily heat energy requirement varies between 0 and 1.3 kWh/m^{2}day depending on the desired greenhouse temperature. A similar change is observed during hours of the day. Heat power requirement in plastic greenhouses installed in the Mediterranean region from 04:00 to 07:00 in January is 85 W/m^2, while after 08:00 this value drops to 0 W/m^2 [8].

Heat power iterations required for a single pane PE plastic greenhouse in Antalya climatic conditions with temperature kept at 14, 16 and 18°C are given in **Figure 3**. As can be seen from the figure, there is need for high heat power only during a very short period of the year. In Mediterranean climatic conditions, 3012 h of heating is required when day/night temperature is kept at 18°C. When temperature is dropped to 16°C, 2567 h of heating is required.

Heat energy values required throughout the production period for greenhouses in different cities in the Mediterranean region are given in **Table 7**. These values are for greenhouses with different equipment and when day/night temperature is kept at 16/18°C and ventilation temperature is kept at 25°C [8]. As can be seen from the table, for each type of greenhouse equipment, the lowest heat energy requirement is observed in Mersin (36°:48′ N) with 72.7 kWh/m^{2}a, while the highest heat energy requirement is observed in Muğla with 198.3 kWh/m^{2}a.

In the Mediterranean region, greenhouses with low technology are not heated. On days when the temperature is low, plants are protected from frost with the help of simple heating stoves.

http://dx.doi.org/10.5772/intechopen.72367

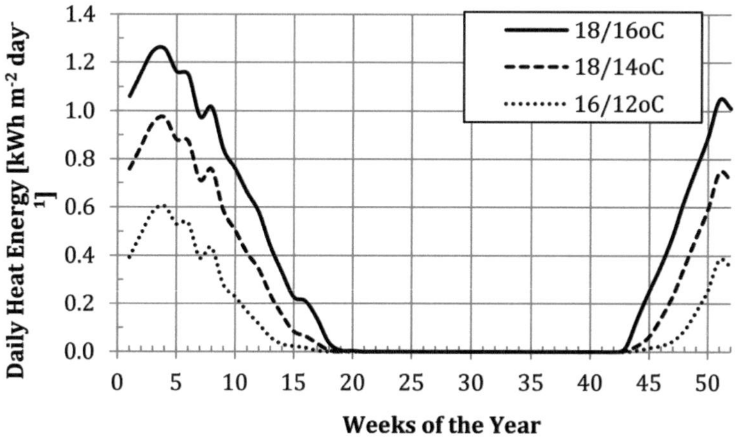

Figure 2. Daily heat energy requirements depending on different day/night temperature values for a single pane PE plastic-covered greenhouse in Antalya climatic conditions (kWh/m^{2}day).

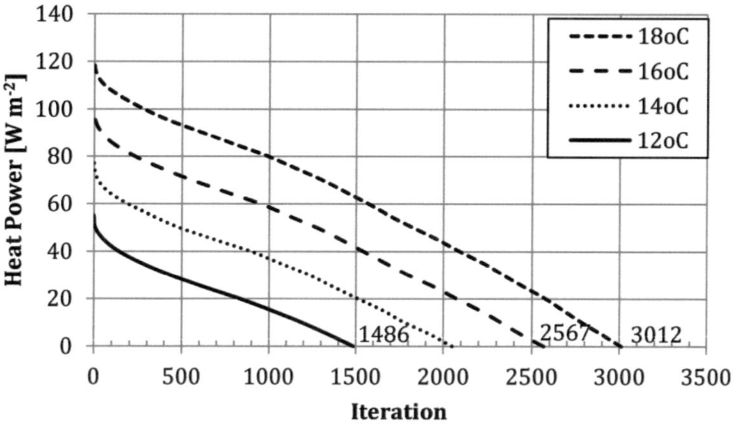

Figure 3. Heat power iterations for different greenhouse temperatures in Antalya climatic conditions.

Cover material	Outside temperature (°C)	Under curtain temperature (°C)	Over curtain temperature (°C)	($t_{i\text{-palt}}-t_o$) (°C)	($t_{i\text{-püst}}-t_o$) (°C)
Plastic	9.1	10.7	8.8	1.6	−0.3
	5.0	5.6	3.3	0.6	−1.7
	8.5	10.0	7.8	1.5	−0.7
Glass	9.0	12.7	9.5	3.6	0.5
	8.2	12.0	8.6	3.7	0.4
	10.1	13.6	10.7	3.4	0.6

Table 7. Under curtain, over curtain and outside temperature values in unheated plastic and glass greenhouses with thermal curtains opened and closed (time 17:00–07:00 Average values).

However, with this kind of heating, heat energy is not distributed properly in the greenhouse, and plants close to the heating stove are harmed. In small family businesses, pipe heating systems are not economical due to greenhouse sizes and initial investment costs. In these businesses, instead of pipe heating, low-cost direct-fire air blast heating systems are preferred.

2.2. Heat energy conservation

As much as greenhouse heating, conservation of heat energy has great importance due to increasing energy prices and CO_2 releases of energy sources. In order to conserve heat energy in greenhouses, multipane cover materials may be used. However, besides temperature and humidity values, light (PAR) in the greenhouse should be kept at the highest levels as it is one of the most significant factors for plant growth. For these reasons, it is suggested to cover the side walls of greenhouse in the Mediterranean region with multipane cover material for heat conservation, while the roof area should be covered with single pane material in order to allow sufficient light into the greenhouse.

Thermal curtains are used for heat conservation in greenhouses. Average under curtain, over curtain and outside temperatures measured in unheated glass and PE plastic Mediterranean greenhouses with open and closed thermal curtains between 17:00 and 7:00 are given in **Table 7**. As can be seen from the table, under curtain temperature in the PE plastic greenhouse with thermal curtains closed is above outside temperature, while over curtain temperature values are below outside temperature. In the glass greenhouse, both under curtain and over curtain temperature values recorded are above outside [7, 9].

Depending on the properties of thermal curtains, heat conservation in heated greenhouses can be achieved at different ratios. In **Figure 4**, fuel quantities based on temperature differences (ΔT) in a PE plastic greenhouse with and without thermal curtains are given [7]. When the thermal curtain is open, the amount of fuel (diesel) required for a 7 K temperature difference is

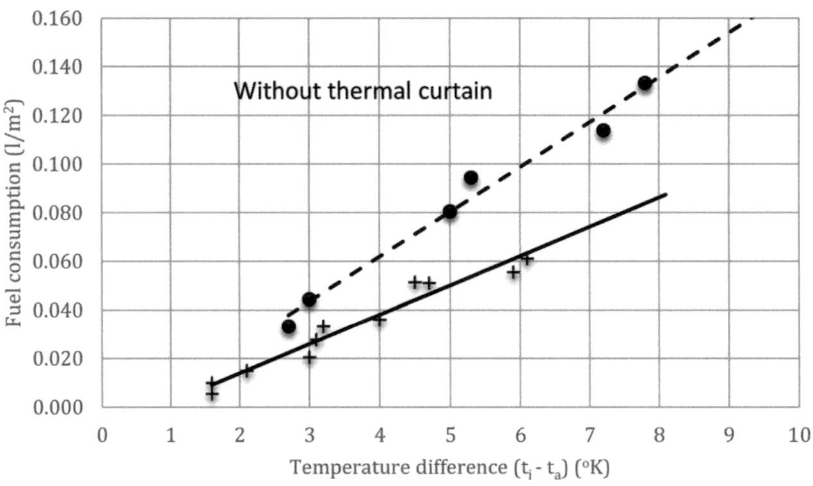

Figure 4. Fuel consumption based on temperature differences (ΔT) in a PE plastic greenhouse heated with direct-fired air blast heating system, with thermal curtains opened and closed.

0.118 l/m^2. The fuel requirement for the same temperature difference with the thermal curtain closed is 0.074 l/m^2. This is equivalent to a 37% energy saving in a greenhouse with a direct-fired air blast heating system.

Impermeability of thermal curtains used in greenhouses is very important in terms of energy savings. Edges where thermal curtains meet the side walls and facades should be leakproof. Otherwise, transfer of heated and rising air through the roof cover material will be unavoidable. Annual heat energy and saving ratios calculated based on the impermeability of thermal curtains in PE plastic greenhouses under Mediterranean climatic conditions with night/day temperature 16/18°C and ventilation temperature 25°C are given in **Table 8** [9]. As can be seen from the table, there will be a 27% difference in energy savings between thermal curtains with perfect insulation and those with poor insulation.

2.3. Passively benefiting from solar energy in unheated greenhouses

In Mediterranean climatic conditions, there is no need for heating during day hours as solar energy reaching the greenhouse exceedingly meets the daily energy requirement of the greenhouse. However, after sunset, greenhouse temperature drops rapidly depending on the thermal properties of the cover material. In a study aiming to passively benefit from solar energy in greenhouses with low technology, water mattresses consisting of transparent PE tubes (with a diameter of 31.8 cm, width of 150 μm and water capacity of 80 l/m) were placed between plant rows. Measurements showed that in the case of using water mattresses in a glass greenhouse, the temperature difference is 2.8–3.4°C, while in a greenhouse without water mattresses, this value is 1.2–2.7°C (**Table 9**) [7].

While the temperature difference in a glass greenhouse with thermal curtains and water mattresses ranges from 6.3 to 8.1°K, the temperature difference in a greenhouse with thermal curtains but without water mattresses were recorded as 1.6 to 2.2°K (**Table 10**) [7].

2.4. Ventilation

In the Mediterranean region, it is necessary to ventilate greenhouses during day hours of the winter months. Ventilation in winter months is done more to regulate CO_2 concentration than

Greenhouse equipment	Heat energy requirement [kWh/m^{2}a]			Savings ratio [%]	
	Impermeability				
	Good	Average	Poor	Good-Average	Good-Poor
Without thermal curtain	118.3				
With thermal curtain	80.5	95.1	109.6	15.4	26.6
Savings ratio	32.0	19.6	7.4		

Table 8. Heat energy requirements based on the impermeability of thermal curtains in PE plastic greenhouses with single pane roof and double pane side walls under Antalya climatic conditions with night/day temperature 16/18°C [7].

$t_{o,min}$(°C)	$t_{i,min-tube}$ With water mattresses (°C)	$t_{i,min-tubeless}$ Without water mattresses (°C)	ΔT With water mattresses (°C)	ΔT Without water mattresses(°C)	$t_{tube}-t_{tubeless}$ (°C)
2.9	6.1	4.1	3.2	1.2	2.0
3.0	5.8	4.7	2.8	1.7	1.1
4.1	7.5	6.8	3.4	2.7	1.7

Table 9. Minimum temperature differences obtained in an unheated Mediterranean greenhouse with water mattresses (°C).

$t_{o,min}$(°C)	$t_{i,min-tube}$ With water mattresses (°C)	$t_{i,min-tubeless}$ With water mattresses (°C)	ΔT With water mattresses (°C)	ΔT Without water mattresses (°C)	$t_{tube}-t_{tubeless}$ (°C)
0.0	6.8	2.2	6.8	2.2	4.6
1.8	8.1	3.5	6.3	1.7	4.6
5.2	11.6	6.8	6.4	1.6	4.8
−0.1	8.0	2.0	8.1	2.1	6.0

Table 10. Minimum temperature differences obtained in an unheated Mediterranean greenhouse with thermal curtains and water mattresses (°C).

to send away high temperatures. It is only possible to obtain the temperatures that plants have adapted to (17–27°C) by regular ventilation from mid-February until the first week of May [5]. In the evaluations based on long year hourly temperature data for Antalya (36°:53′ N), where greenhousing is a common agricultural practice, it is seen that the temperature is above 26°C in 1628 h of the year (**Table 11**).

Temperature iterations for temperatures above 26°C based on the ratio of ventilation openings to greenhouse floor area are given in **Table 10**. As can be seen from the table, as the ratio of ventilation openings to greenhouse floor area increases, iterations for temperatures above 26°C decrease. In Antalya climatic conditions, when the ratio of ventilation openings to greenhouse floor area is 20%, during 206 h of the total 744 h of May, the greenhouse temperature is above 26°C.

Taking into consideration the long year hourly climatic values of Antalya, hourly temperature values calculated for greenhouses with different ventilation openings are given in **Table 12**. As can be seen from the table, average outside temperature values obtained from long year climatic values in May vary between 16 and 26°C. The simulation calculations show that when the outside temperature is 25.7°C at 12:00 in May, temperature in a greenhouse with 5% ventilation opening is 30.4°C and temperature in a greenhouse with 10% ventilation opening is 28.5°C.

2.5. Ventilation and shading

Starting from the first week of May, greenhouses in the Mediterranean region are shaded with clay or whitewash. With shading, greenhouse temperature rises are prevented by reducing solar radiation that reaches the greenhouse. In greenhouses installed in recent years, solar radiation is reduced by partially opening thermal curtains. In June, under Antalya climatic conditions, when radiation reaching the greenhouse is reduced by 50% with shading and when the ratio of ventilation openings is 20%, greenhouse temperature is above 26°C for 96 h.

A_V/A_G	January	February	Mach	April	May	June	July	Augusts	September	October	November	December	Total
Number of hours when outside temperature is above 26°C													
	0	0	0	0	0	311	487	466	298	66	0	0	1628
Number of hours when greenhouse temperature is above 26°													
0.01	102	145	237	290	366	405	518	488	357	288	200	105	3501
0.05	0	0	78	205	335	389	504	480	336	254	98	0	2679
0.10	0	0	0	82	278	382	496	479	329	222	40	0	2308
0.15	0	0	0	18	246	375	489	477	323	201	10	0	2139
0.20	0	0	0	5	206	373	488	476	321	177	2	0	2048
0.25	0	0	0	0	160	370	488	474	320	163	0	0	1975
0.30	0	0	0	0	130	367	488	474	319	148	0	0	1926
0.35	0	0	0	0	113	362	487	474	318	142	0	0	1896
0.40	0	0	0	0	99	357	487	474	317	134	0	0	1868

Table 11. Temperature iterations for temperatures above 26°C outside the greenhouse and in a plastic greenhouse with different ventilation opening ratios under Antalya climatic conditions in May (h).

Time	Ratio of ventilation opening ratio to greenhouse floor area ($\frac{A_V}{A_G}$) (%)					Outside temperature t_a (°C)
	1	5	10	15	25	
	Temperature in the greenhouse t_i (°C)					
7	23.1	21.7	21.1	20.8	20.6	20.0
8	28.0	25.5	24.4	23.9	23.4	22.4
9	32.0	28.4	26.9	26.2	25.6	24.2
10	34.6	30.0	28.2	27.4	26.6	25.3
11	35.7	30.4	28.5	27.7	27.0	25.6
12	36.1	30.4	28.5	27.7	27.0	25.7
13	35.7	30.1	28.3	27.5	26.8	25.6
14	34.8	29.6	27.9	27.2	26.6	25.5
15	33.1	28.7	27.2	26.6	26.1	25.1
16	30.5	27.3	26.2	25.7	25.3	24.5
17	28.0	25.9	25.1	24.8	24.5	23.9
18	25.0	24.0	23.6	23.4	23.2	22.9
19	22.0	21.8	21.8	21.7	21.7	21.6

Table 12. Inside temperature values calculated for a plastic greenhouse with different ventilation opening ratios (A_V/A_G) under Antalya climatic conditions in May.

Time	Ratio of ventilation opening ratio to greenhouse floor area ($\frac{A_V}{A_G}$) (%)					Outside temperature t_a (°C)
	1	10	15	20	25	
	Temperature in the greenhouse t_i (°C)					
7	27.0	25.8	25.6	25.5	25.5	25.1
8	30.3	28.3	28.0	27.9	27.8	27.3
9	32.8	30.1	29.7	29.5	29.4	28.8
10	34.5	31.1	30.7	30.5	30.3	29.6
11	35.4	31.5	31.1	30.8	30.6	29.9
12	35.6	31.6	31.2	31.0	30.8	30.1
13	35.1	31.2	30.8	30.6	30.4	29.8
14	34.5	30.8	30.5	30.3	30.1	29.6
15	*33.6*	*30.5*	*30.2*	*30.0*	*29.9*	*29.4*
16	*32.3*	*29.9*	*29.6*	*29.5*	*29.4*	*29*
17	*30.7*	*29.0*	*28.8*	*28.7*	*28.7*	*28.3*
18	*28.9*	27.9	27.8	27.7	27.7	27.5
19	26.5	26.3	26.2	26.2	26.2	26.1

Table 13. Temperature values calculated for different ratios of ventilation openings in a plastic greenhouse where 50% shading is done in June, under Antalya climatic conditions.

Although solar radiation causing increases in greenhouse temperatures under Mediterranean climatic conditions can be reduced to a certain degree with shading, it is not possible to obtain the environment temperature that can be tolerated by plants using shading in certain months of the year. Temperature values based on hours of the day and ratios of ventilation openings for a plastic greenhouse where 50% shading is done during June are given in **Table 13**. As can be seen from the table, even when the ratio of ventilation openings to greenhouse floor area is 25%, temperature values in a shaded greenhouse are above 30°C at 10.00–14.00. Under these conditions, evaporative cooling becomes necessary for continuation of plant growth.

2.6. Cooling

When average daily temperature is above 22°C and the maximum temperature is 27°C, active cooling in the greenhouse is necessary [10]. When average daily temperature values of the Mediterranean climate zone are reviewed, it is seen that starting from June average temperature values are above 22°C (**Table 3**).

Temperature and humidity values calculated in August for a glass greenhouse in Adana (37°:01′ N) with shading and evaporative (Fan&Pad) system is given in **Figure 5**. As can be seen from the figure, in the Mediterranean region, outside temperature values during the day can be as high as 35.2°C in August. Despite shading and evaporative cooling, temperature in the greenhouse reached 29.1°C at 15:00. As can be seen from the figure, humidity values throughout the day varied between 90 and 98%. In Adana climatic conditions, using shading and active cooling in August resulted in a temperature difference (ΔT) of 6.1 K [11].

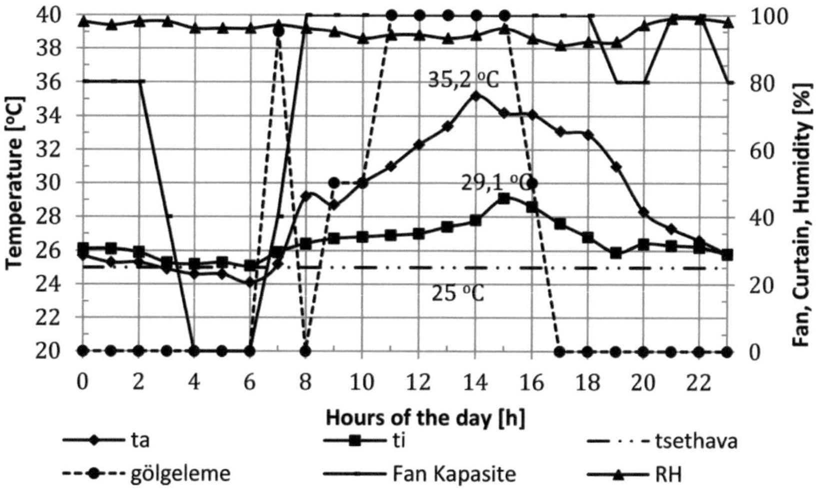

Figure 5. Temperature values calculated in August for a glass greenhouse obtained with shading and evaporative cooling.

3. Result and evaluation

In order to be able to achieve high-quality production and efficiency in the Mediterranean climate zone, greenhouses should be heated during November–March, ventilated and shaded during February–May and cooled during June–September. On the Mediterranean coastline, simple greenhouses are not heated regularly as average daily temperature does not fall below 7°C. In such businesses, when the temperature is very low, plants in the greenhouse are protected against frost using simple methods. Depending on the production (single production, spring or fall production), a 9–12 kg/m^2 efficiency can be obtained in tomato production. In simple plastic greenhouses, it is not economical to install pipe heating systems due to greenhouse dimensions and cost of heating systems. In these greenhouses, instead of pipe heating systems, direct-fire air blast heating systems could be used. However, in this case, it is necessary to choose cheap fuel as well.

In greenhouses built on the Mediterranean coastline in recent years, central heating systems are installed, and coal is used as fuel. Heat requirement of a single pane PE plastic greenhouse on the Mediterranean coastline, in which there is no heat conservation and day/night temperature is kept at 18/16°C day/night, varies between 95 and 256 kWh/m^2 depending on the climatic properties of the greenhouse site. This is equivalent to approximately 18–47 kg/m^2 year imported coal.

As much as greenhouse heating, heat conservation in heated greenhouses has great importance. Since total solar radiation in the Mediterranean region is lower than 2.34 kWh/m^{2}day during December and January, greenhouse roofs are covered with single pane and side walls are covered with double pane cover material. In the Mediterranean region, thermal curtains are used in greenhouses for heat conservation. In Antalya climatic conditions, heat requirement for a PE plastic greenhouse with side walls covered with double pane thermal curtains is 80.5 kWh/m^2 area when night/day temperature is kept at 16/18°C. In a greenhouse without thermal curtains, this value is 118.3 kHz/m^2 areas. In other words, 32% heat energy is saved in a greenhouse with thermal curtains. This is equivalent to 7.1 kg/m^2 year imported coals.

In Mediterranean climatic conditions, 30 kg/m^2 truss tomatoes can be produced in heated modern greenhouses which in heated greenhouses where CO_2 fertilization is done, the efficiency is 40 kg/m^2. Approximately 18–28 kg/m^2 efficiency increase in heated greenhouses should cover heating expenses. Depending on the climate of the region, the cost of truss tomato production in heated modern greenhouses varies between 1.29 and 1.69 TL/kg. In the feasibility calculations made for modern greenhouses in the Mediterranean region, return on investment is 14–25%, depending on the production methods of the business [12]. The quality of products obtained from heated greenhouses is higher than the quality of plants grown in unheated greenhouse. Also, due to humidity control, agricultural pesticide use is less.

In order to benefit more from solar energy on the Mediterranean climate zone, it is appropriate to use water mattresses. With the help of water mattresses, a 2–3°K temperature difference can be obtained, but when water mattresses are used together with thermal curtains, the temperature difference obtained becomes 6°K. However, using water mattresses, it is not possible to obtain the optimum greenhouse temperature during night hours.

In Mediterranean climatic conditions, a 20–25% ratio for ventilation openings in the roof area is sufficient. Increasing the sizes of the ventilation openings on the roof in May has very little impact on temperature difference. However, it should be kept in mind that insect screens placed on ventilation openings in modern greenhouses decrease the effectiveness of ventilation.

Greenhouse shading implemented in a way that it does not affect air circulation can help reduce temperature difference in the greenhouse when used together with ventilation. In a greenhouse with 20% ventilation area on the roof, reducing solar radiation by 50% with shading can reduce the temperature difference (ΔT) in June up to 1°C. However, in June, evaporative cooling is needed as outside temperature around noon is above 30°C. In Mediterranean climatic conditions, temperature difference obtained with evaporative cooling in August (depending on outside humidity and exchanged air volume) is nearly 6°C. However, evaporative cooling is not preferred in production greenhouses because it requires electrical energy and a high quantity of clean water. For this reason, greenhouses on the Mediterranean coastline should be left idle starting from the second week of June.

Author details

A. Nafi Baytorun[1] and Zeynep Zaimoglu[2*]

*Address all correspondence to: zeynepzaimoglu6@gmail.com

1 Department of Agricultural Structures and Irrigation, Çukurova University Faculty of Agriculture, Balcalı-Adana, Turkey

2 Department of Environmental Engineering, Çukurova University Faculty of Engineering and Architecture, Balcalı-Adana, Turkey

References

[1] Munoz P, Anton A, Nunez M, Paranjpe A, Arino T, Castells X, Montero JI, Rieradevall J. Comparing the environmental impacts of Greenhouse versus open-field tomato production in the mediterranean region. Acta Horticulturea. 2008;**801**(2):1591-1596

[2] Giuliano G, Gordon P, Pan Q, Park JY, Wang LL. Estimating freight flows for metropolitan area highway networks using secondary data sources. Networks and Spatial Economics. 2010;**10**(1):73-91

[3] TÜİK. Tarım istatistikleri özeti. Ankara-Türkiye: Türkiye İstatistik Kurumu; 2014

[4] Zabeltitz Chr.von. Integrated Greenhouse Systems for Mild Climates. Springer: Verlag Berlin Heidelberg. 2011

[5] Eker MM. Jeotermal seralarda hedef, 30 bin hektar. Jeotermal belediyeler dergisi. sayı 6. 2012: s.5-14

[6] Nisen A, Grafiadellis M, Jiménez R, La Malfa G, Martinez-Garcia PF, Monteiro A, Verlodt H, Villele O, Cv Z, Jc D, Baudoin W, Jc G. Cultures protegees en climat mediterranean. Rome: FAO; 1988

[7] Baytorun AN, Abak K, Tokgöz H, Güler Y, Üstün S. Seraların kışın iklimlendirilmesi ve denetimi üzerinde araştırmalar. Türkiye Bilimsel ve Teknik Araştırma Kurumu. Proje no TOAG-993; 1995

[8] Baytorun AN, Akyüz A, Üstün S. Seralarda ısıtma sistemlerinin modellemesi ve karar verme aşamasında bilimsel verilere dayalı uzman sistemin geliştirilmesi. TÜBİTAK Proje No: 114O533; 2016

[9] Baytorun AN, Abak K, Daşgan HY, Topçu S. Climatic problems of the plastic greenhouses in Turkey. International Congress for Plastics in Agriculture. CIPA Procedings. Tel Aviv. Israel; 1997

[10] Kittas C, Katsoulas N, Bartzanas T, Bakker S. Greenhouse climate control and energy use. Good Agricultural Practices for greenhouse vegetable crops. Principles for Mediterranean Climate Areas. Rome: FAO; 2013

[11] Tekinel O, Baytorun N, Demir Y. Çukurova koşullarında seralarda ıslak yastıklarla soğutma olanakları. Türkish Journal of Agriculture and Forestry. 1989. Cilt:13, Sayı 3b, s: 1284-1293

[12] Baytorun AN. Sera projelerinin hazırlanmasında dikkate alınacak kriterler, ve TR 63 bölgesi için örnek sera fizibiliteleri. DOĞAKA (Basımda); 2016